Bath In The Eighteenth Century: Its Progress And Life Described

William Tyte

BATH

IN THE

EIGHTEENTH CENTURY:

ITS

PROGRESS AND LIFE DESCRIBED,

BY

WILLIAM TYTE,

(Author of a " History of Lyncombe and Widcombe,' Etc.)

BATH:

"Chronicle" Office, Kingston Buildings;

G. & F. Pickering, Booksellers, Bridge Street.

1903.

JOHN WOOD THE ELDER.

(See Note at back of Title).

PREFACE.

— -

The idea of writing a History of Bath in the Eighteenth Century is by no means a new one. Three or four citizens in the past felt the need of such a work and gathered materials for its execution. In neither instance were the limits of this preparatory stage exceeded, although the ability for its satisfactory completion was not lacking. That there was no fruition is therefore to be regretted, especially as it would have shielded me from the temptation of trying to do what others had left undone. Their intentions however prove, as above-mentioned, that a work of the kind in question was desirable, and, taking the same view, it seemed to me that longer inaction might lead to permanent failure. Having myself collected a varied stock of memoranda relating to our olden times, and the gleanings by another having come into my possession, I, like my predecessors, long dallied with the notion of squeezing the mass of notes into a book. Hesitation was partly due to the difficulty of knowing what plan to adopt—whether to write a general history of the period, or divide it into epochs and deal with each separately. The former, I feared, would be too sketchy, the latter too disjointed. Not liking either, I at length resolved to digest the materials into subjects and let each run the length of the century, as far as it admitted of this extension. If the arrangement adopted should not be the best, some other toiler in the same literary field may be able to improve upon it hereafter. To this explanation of its genesis, I have only to add the hope that the volume may not be deemed either unworthy the labour bestowed upon it or a superfluous addition to Bath bibliography.

THE AUTHOR.

Bloomfield Crescent, Bath, March, 1903.

CONTENTS.

BATH

IN THE

EIGHTEENTH CENTURY.

INTRODUCTION.

THE most interesting and important era of Bath is the eighteenth century. Whatever reputation it enjoys, other than that derived from its mineral springs and natural scenery, is the outcome of the development which then took place. Its growth was not more remarkable than the dignity and beauty stamped upon its buildings. Homes suitable for the élite of society were provided in the suburbs of the old town, and were rapidly occupied by the class for whom they were intended. This expansion necessarily revolutionised the life of the city. Instead of the local prosperity depending on casual visitors, who came for health or pleasure, it was based now on the residents of wealth and distinction included in its permanent population. Concurrently with this change in the external aspect of the town, care was taken to provide amusements adapted to the taste of the age, while a code of rules was laid down to ensure order and decent behaviour in places of public resort. Hence a social state quite unique grew up, not to be personally familiar with which was deemed to be without Fashion's imprimatur. Glittering as it was on the surface, there was much that was objectionable seething beneath. The coarseness and profligacy of the age, together with its sordidness, leavened the mass; the evil was the more baneful because vice in its many forms was tolerated as a necessity. It is this mixture of the gay and the sombre that gives to Bath at this period the same fascination to authors as the Court of France does during the reigns of Louis XIV. and Louis XV., the debaucheries of which were gilded by a grace and stately etiquette that imparts the charm so many find irresistible. Without having the piquant personal memoirs which illumine the " ancien régime " of our neighbours, enough is known of contemporary life in Bath, both from the records extant and the pages of fiction, to make it familiar to most readers. None the less there is room for a work dealing with the century as a whole, wherein local material progress and social conditions may be depicted on a wider canvas than hitherto has been attempted.

At the commencement of the century the city was so small, its accommodation so inferior, and the Baths so uninviting, that it is not easy to understand its claim to be regarded as a place of fashionable resort. In Coxe's "Magna Britannia et Hibernia," published in 1720, a very good, if flattering, picture of it is given: "The city of Bath (it says) is but a small city, but very compact, and well inhabited by means of the medicinal waters. No man can imagine otherwise how so great a company frequents it for three parts of the year; that it is said there are usually there eight thousand families at a time (!)—some for the benefit of drinking the waters, others for bathing, others for diversion and pleasure, of which last there is no place in Europe that affords more. It is walled round with a high stone wall pretty entire, and hath a street built upon it from whence there is a pleasant prospect over the meadows on the west side. It has four gates, viz., (1) Northgate, with its suburbs, leading to London, opens into High Street, where there is a plentiful market, kept under the Town or Council Hall, a neat building standing upon twenty-one pillars; in front are the effigies of two Kings—Coel, who is said to have given a charter to this city, and Edgar, a Saxon, who was crowned here anno. 973. From this place the street dividing leads to (2) Westgate, a handsome building of stone, and containing some of the best apartments in the town. The other street leads to (3) Southgate, from thence through suburbs to the bridge over the Avon, in the middle of which is an old gateway. The other (4) gate, to the East leads to the river, where there is a ferry. The streets are narrow, but well paved, and the buildings, by reason of the great plenty of stone thereabout, extraordinary neat, some of them as handsome stone buildings as are anywhere found; but many of them standing in courts and alleys, where coaches can't go. There are forty Sedan chairs licensed by the Mayor, which, for sixpence, are obliged to carry a person from any one part of the city to the other within the walls. There is also another good regulation which has much tended to the benefit of the town, that no person shall demand above ten shillings per week for one room, which, freeing men from such impositions as are common in other places of concourse, hath brought such numbers of people to it, more for diversion and drinking the waters, that the citizens have been forced to erect many new buildings, yea, whole streets, for the accommodation of strangers, viz., in the north suburbs and without Westgate. Adjoining to the wall on the north side of the town there has lately been raised a neat stone building for a school house, which was erected and is now maintained by the contributions of strangers who come to the waters. To allure these there is nothing wanting that may please or divert, for here is a little Theatre; pleasant walks upon the town walls, thronged every evening with the most agreeable of both sexes, and along the side raffling shops; and adjoining to the walls, without Mr. Harrison's house, there is a fine ballroom and pleasant gardens down to the river. Besides all these things, the neighbouring hills afford the most pleasant down imaginable, where it is incredible what a number of coaches and horses appear there at a race, so that Bath is one of the most proper places in the world either for obtaining or preserving health by that constant cheerfulness which the agreeable company and virtue of the waters infuse into them that go thither."

It is difficult to believe that the city at this early period was accustomed to have "at one time" as many as eight thousand visitors. Such an influx may have occurred during the visit of Queen Anne, and it is probable that this exceptional aggregate came to be regarded as the normal number. Eight thousand would be an unduly large estimate for the whole season, looking at the limited number of houses in existence, including, too, the new ones to which reference is made. Even as late as 1727 the city comprised no more than fifteen streets, sixteen lanes, four inferior courts, five open areas, four terrace walks, three alleys, four throngs, and a few private courts. Without accepting the figures mentioned, it may be safe to assume that the strangers flocked to the city in sufficient force to put to a severe test the accommodation obtainable. Such patronage was only the natural result of the reputation Bath had gained as a health resort, and the home of "cheerfulness" and "agreeable company." The prime factor in its prosperity was, no doubt, its healing springs, which despite all the changes—dynastic, social, and political—never lost their vogue from Roman times onward. People were content to put up with a jumble of houses and ill-furnished apartments because the standard of taste and comfort in their own homes was not high.

A great deterrent to dwellers at a distance was the state of the roads, which rendered travelling a weariness to the flesh. Narrow and full of ruts and holes as they were, a journey hither must have been like a penitential pilgrimage. Breakdowns were common, stoppages from sinking into quagmires had often to be faced, nor was it at all unusual for an ill-defined road to be forsaken through ignorance of its direction. As an illustration of the dangers besetting travellers an incident that occurred to Sir Richard Steele may be mentioned. He was returning to the city with Mr. (afterwards Alderman) Bush from Claverton Down, when, noticing a large cavity in the road, he suggested to his companion that he should dismount and fill it with some stones near, which was done, Sir Richard remarking, "Young man, you have the satisfaction to reflect that you may save the life or the limbs of your fellow creatures." If the highways were bad, the waterway was no better. Though the Avon flowed round the city, it was useless for traffic beyond owing to the obstruction caused by successive weirs and occasional shallowness. It was not till the second decade of the century that the river was rendered navigable to Bristol, and not till the third that canalisation was completed to make the passage of boats safe and easy.

With all these disadvantages, visitors, as we have seen, were never lacking, nor given to grumbling at their quarters. It mattered not to them that the city consisted of groups or rows of houses crowded round the Abbey, with the Baths intermixed, or that mediæval walls—a sure sign of stagnation—surrounded it. The whole extent was, in fact, comprised in an area, bounded by the Upper and Lower Borough Walls (as now known) on the north and south, and the Sawclose and the Orange Grove on the west and east, the only extension being Trim Street on the north side (built in 1707 by Mr. Trim, a wealthy clothier, and named after him), and a few houses on the west side. It may be that the lack of increase in the population was partly due to the high death-rate, from the insanitary condition under which people lived, the outbreak of epidemics and the empirical treatment of disease by the faculty. In 1700 it is computed that the number of householders was about three hundred, and the resident population not more than two thousand. The fluctations were wide from the coming and going of strangers.

How all these were housed before Wood began his extensive building operations is somewhat of a mystery. Overcrowding must have been common, even among the élite, and that was a matter of which no by-law took cognisance. For Royalty, apartments were provided over the Westgate, or at the Priory House near the Baths. Princess Amelia on her first two visits (1728 and 1732) lodged in a mansion in Broad Street, as did the Prince of Orange in 1734. The large gabled houses surrounding the King's and Queen's Bath, tenanted chiefly by members of the Corporation, received other persons of rank and distinction. There were also second and third-rate lodging-houses, and the several hostelries, where the less opulent could always find quarters, if the pressure on them was not too great. Of the surroundings one and all offered to their guests, Wood, the architect, has left a vivid description. Speaking of his early recollections he says: "About this time (1725) the boards of the dining rooms and most other floors were made of a brown colour with soot and small beer, to hide the dirt as well as their own imperfections, and if the walls of any of the rooms were covered with wainscot, it was such as was mean and never painted. . . . With cane or rush-bottomed chairs the principal rooms were furnished, nor were the tables or chests of drawers better in their kind; the looking-glasses were small, mean, and few in number; with Kidderminster stuff, or at best with cheyne, the woollen furniture of the principal rooms was made; and such as was of linnen consisted either of corded dimaty or coarse fustian; the matrons of the city, their daughters, and their maids flowering the latter with worsted during the intervals between the seasons, to give the beds a gaudy look."

From the foregoing it will be seen what scope there was for brightening and refining the internal equipment of the dwellings, as well as for constructing others in better situations and more commodious for the tenants. Whence came the stimulus to this reforming work, and how triumphantly it was carried through, the ensuing pages will disclose.

CHAPTER I.

THE BUILDERS OF BATH.

JOHN WOOD THE ELDER.

" I monumentum requiris circumspice "—the motto in St. Paul's Cathedral identifying Sir Christopher Wren with this his masterpiece—may well be applied to the two Woods and their works in Bath. The monument of each is to be found in the grand parades, crescents, squares, and other public buildings with which they adorned the city and its suburbs. If they could not say, as Augustus boasted, that he found Rome brick and left it marble, the transformation effected by them here was scarcely less striking. A crowded, smelly, wall-begirt town, with a labyrinth of narrow streets and courts was, by the Woods, and the example set by them to their contemporaries and successors, changed to a beautiful, airy, commodious city, with few others to equal it for its architectural features and the expansive views opened for the enjoyment of residents and visitors.

It is much to be regretted that we know nothing about the parentage, birth-place, or early education of Wood. Were this information available it would be possible to explain how it was that he, when quite young, had so completely mastered his profession, in the classic branch at least, as to be able (in the words of Macaulay) to design buildings rivalling in dignity and beauty the best works of Bramante or Palladio. Wood, when he settled in Bath in 1727, and plunged into his grand building schemes, was only 23 years of age. Seeing what he accomplished in the five-and-twenty years of life he afterwards enjoyed, his rapidity in planning must have been accompanied by an industry truly prodigious. At the same time he had no end of difficulties to overcome in securing the land he required for carrying out his projects, and in getting contractors to adhere to his rules. Again and again he was incensed to find that the freestone used was damaged, the edges not straight, or some of an inferior quality introduced. Nothing short of the best would, in fact, satisfy him; scamping of any kind was his abhorrence. Probably this may help to explain " the world of calumnies, falsehoods, and discouragements " a contemporary notice of him states that he had to endure. Working under these conditions, and with such ceaseless industry, it is not surprising that he died in his prime.

A man so remarkable naturally arouses curiosity as to his early history and the kind of training he received. As already stated, it is not possible now to gratify these yearnings. He himself is silent on the subject; there is not a trace of the autobiographical in any of his published writings. How he came to select Bath as a field for his operations can likewise only be conjectured. The probability is that Ralph Allen, when journeying through the country to establish his system of cross posts, and to find capable and honest men for his staff, met Wood in some Yorkshire town, and being struck by his skill with his pencil and his intelligence, suggested that he should try his fortunes in the western city, where there was a good opening for an architect of ability. It is no reflection on Allen to say that he was not wholly disinterested in giving this advice. He was just maturing his plans for embarking in the stone trade at Combe Down. He knew the great want of suitable house accommodation for visitors and the fine sites available. An architect with enterprise, large views, and talent to carry these out, would create a capital local market for the stone locked in the bowels of the earth, while an external demand was expected from the improvement of the Avon, just rendered navigable to Bristol. The advantages in prospect were therefore reciprocal, as the sequel showed. Wood, who was in his twenty-first year, entered into the scheme with eagerness. In the summer of 1725 a plan of Bath was sent to his home in Yorkshire (presumedly by Allen) and he at once set about devising methods for the improvement of the town. His attention was fixed on two areas, one on the north-west corner, the other on the north-east (Bathwick) side— the former belonging to Mr. Gay and the latter to Lord Essex. With Mr. Gay, who was an eminent surgeon in Hatton Garden, London, and the owner of the estate now occupied by Queen Square, Gay Street, the Circus, Brock Street, and the Royal Crescent, Wood had an interview on the last day of December, 1725, and communicated to him a design, as he did one to Lord Essex in March following. It is interesting to learn what the youthful improver thought were the requirements of the city. He says: "I proposed in each design to make a grand place of assembly, to be called the Royal Forum of Bath, another place no less magnificent for the exhibition of sports, to be called the Grand Circus; and a third place, of equal state with either of the former, for the practice of medicinal exercises, to be called the Imperial Gymnasium, from a work of that kind taking its rise first in Bath during the time of the Roman Emperors." That such novel schemes did not find favour with the owners of the land is not astonishing. Something more practical was needed, and, having revised his plans, Wood came to an agreement with Mr. Gay in November following.

Finding that experienced workmen would be needed, he to get them turned builder himself, and commenced operations in what is known as Chandos Court, which he built, as well as adjacent property, for the Duke of Chandos. At the same time he contracted to make a canal to complete the navigation of the Avon between Bath and Bristol. In connection with these works he makes some rather startling revelations as to the condition of the "skilled labour market." For the better execution of the canal, he procured labourers who had been employed on the Chelsea Water Works, and sent them down to Bath, "till which time the real use of the spade was unknown in or about the city, and the removal of earth was then reduced to a third part of what it formerly cost." He adds: "I likewise provided masons in Yorkshire, carpenters, joiners, and plasterers in London and other places, and from time to time sent such as were necessary down to Bath to carry on the buildings I had undertaken. And it was then, and not till then, that the lever, the pulley, and the windlass were introduced among the artificers in the upper part of Somersetshire; before which time the masons made use of no other method to hoist up their heavy stones than that of dragging them up with small ropes against the sides of a ladder."

The agreement with Mr. Gay was, however, shelved, through that gentleman's reluctance to offend the Corporation. He was one of the representatives in Parliament, and knew full well how averse the Town Council, by whom he was chosen, were to anything being erected outside the borough walls. Hence his hesitation. The delay was a great discouragement to Wood, who at once made a move, strikingly illustrative of the boldness of his genius. He conceived the idea of rebuilding the town, and prepared the necessary plans, which he submitted to the Corporation, who "thought proper (he says) to treat all my schemes as chimerical." Baffled here, he turned again to Mr. Gay, from whom he leased ground sufficient for the building of the fourth part of a square, to be named after the Queen, and the approaches thereto—Barton Street and John Street. In November, 1728, the foundations were commenced, and in seven years it was completed, except on the west side, besides sixty houses in the streets leading to the Square, together with a chapel (Queen Square Chapel, since removed) for the use of the inhabitants. Meanwhile Wood had negotiated six other leases for land from Mr. Gay; he had also built Dame Lindsay's assembly room and four edifices on the east side of the old bowling green (the south side of the Abbey), likewise adding a wing on its north side, and new fronting Ralph Allen's house as well as raising it a storey higher. This portion of the house, which its architect described as "a sample of the greatest magnificence ever proposed by him for our city houses," still stands amidst squalid surroundings. It is very likely that it was not absent from the mind of the Earl of Burlington when he designed Nassau House in the Grove, recently demolished.

Allen contrived to obtain one-fourth part of the Bowling Green as a garden to his residence, although the concession was considered an invasion of the public rights, for here, close under the walls of the Abbey, the diversions popular in that day were held, henceforth to be discontinued. As Wood gladly records, "smock racing, pig racing, playing at football, and running with the feet in bags, thereby received their final end; nor did grinning, stareing, scolding, eating hot furmity, laughing, whistling, and jiging upon the stage, for rings, shirts, smocks, hats, etc., escape the common ruin, these amusements falling likewise."* Strange as it seems now that such rude sports should have been tolerated in the heart of a place of fashionable resort, it is more strange to note the utter obliviousness of the Corporation to sanitation. The condition of the Grove attests this fact. It was an open area extending to the city wall on the east, and consisted of two levels, with steps to ascend from one to the other. Both sections were allowed to remain in a disgustingly filthy state, any cleansing, except on some special occasion, being deemed a superfluous expense. Wood's description must refer to a time anterior to the planting of rows of trees in the Grove and its becoming a fashionable promenade—a change made many years before Wood's work was published. The Abbey Orchard, not far off, was covered with "old motes and ponds" and was "little better than unfathomable bog." If the "pestilence that walketh in darkness" was not rife, the immunity of the city was probably due to its being encircled by meadows, the pure air from which neutralised the disease-engendering microbes which the internal foulness propagated.

In both these directions Wood was ambitious to effect an improvement. With regard to the Grove, he made a design in 1727, for building a row of handsome houses on the east side, the city wall being retained as an area in front; but he laments that "caprice and ignorance interfering, robbed the city and every party of the glory of that design." At a later period (1739) he obtained a lease of the Abbey Orchard, and forthwith commenced building thereon the Grand Parade—North and South Parades, Duke Street, Pierrepont Street, Galloway Buildings, etc. This expansion, which Wood viewed with pride, was not regarded with the same satisfaction by some residents and visitors. Mrs.

---

*A new bowling green was afterwards made on the north side of the Borough Walls, near where Green Street now stands.

Delany, writing in 1736, says, " This town is grown to such an enormous size that above half the day must be spent in the streets, going from one place to another. I like it every year less and less." Still, she oft returned, and at times with the Dean, her second husband, who derived benefit from the waters.

While busy with his earlier projects, he undertook the post of consulting manager of Allen's stone works—an arrangement that did not improve their cordial relations, as may be inferred from Wood's reference to it. " By the resort of strangers to Bath, the fame of the works (he says) was soon carried to the principal parts of the kingdom, and letters came to him (Allen) in great abundance, some with draughts, with a request to know what would be proper for such and such purposes. To enable him to answer all such letters, together with personal applications to the same effect, he proposed to allow me a certain sum per annum in consideration of my giving him proper instructions, which he accordingly had from me during the course of about five years. But the good consequence to the stone trade from all the information the people were gratified with was trifling in respect to the trouble of composing it, and I have the justest reason to lament that my time had not been better employed." Looking at his activity all the while in his own private affairs, Allen's business could not have absorbed so much of Wood's time as represented; possibly it was the termination of the engagement that left a rankling sore, which was not healed when, in after years, he recorded his grievance. Allen, with his generous disposition and fine sense of honour, is not likely to have treated Wood unjustly. The latter was highly strung and irritable; quick in his perceptions and prompt in action, he was inclined to be dictatorial. With these attributes it was not easy to work comfortably with him, and this partly explains his unpopularity.

Whatever Wood's defects, he is certainly entitled to the credit of being one of the chief promoters of the Royal Mineral Water Hospital, in the establishment of which he took a keen interest. The great difficulty was to obtain a suitable site. Again and again the trustees were disappointed; again and again Wood prepared plans for the fresh situation thought eligible. That success at length crowned these efforts was mainly due to his foresight and tact. The theatre, and property adjoining, near the Borough Walls on the north, were secured, but a stable held by other hands was needed. To obtain it at reasonable cost Wood resorted to a subterfuge. A plan drawn by him for the Hospital was adopted by the trustees, printed, and circulated. It showed that the stable in question was not needed. The architect then resumed negotiations with the owner, got the stable, and purchased it; after that he produced his real plan, which, improving the accommodation of the proposed Hospital, was approved, and ordered to be carried into execution forthwith. Wood never accepted a farthing for his several plans, and superintended the erection of the building without fee or reward. He declares, indeed, that every person concerned in bringing the charity into existence performed his part at his own expense; also that the charge of every meeting relating to it was defrayed by those who were present at it.

Of the several other buildings erected by Wood in and around Bath detailed mention need not be made. Elegant examples of private houses may be seen at Batheaston (where he sometimes resided, and his son after him), at Bathford, and elsewhere. He made the plans for the Circus, which the younger Wood carried out; the Exchange, Bristol, was likewise designed by him, in regard to which a critic some years ago stated that it was Wood's most perfect work: " A Palladian edifice more unexceptionable on the whole than any building by Palladio "—all of whose chief works the writer declared that he had seen. The story of the building of Prior Park is too well known to make its repetition desirable. That Allen vetoed or modified the original designs was one of Wood's deepest mortifications. He had hoped to see "the orders of architecture shine forth in all their glory;" but, the warmth of Allen's resolution abating, "an humble simplicity took their place"—a disparaging comment which the structure, shorn as it may be of the ornate, certainly does not justify—a comment, too, tainted with malevolence, the intention being to wound Allen. Moreover, the epithet, "humble," was probably not without reference to Pope's well-known couplet on Allen; and if so it was deliberately meant to barb the shaft aimed at a friend of long standing.

Among works attributed to Wood is the Grammar School, Broad Street. That he was commissioned by the Corporation, in 1742, to prepare plans is beyond doubt; but it is not so certain that they were carried out, as he himself states that "they proceeded no further than to employ me to make a proper design," for which he was paid. He was also requested to make a general plan of the Baths with a view to their improvement. Of the pressing need of the latter we are not left in any doubt. The slips leading into the King's Bath looked "like cells for the dead;" "dark as dungeons, more fit to fill the bathers with the horrors of death than to impress them with the efficacy of the hot waters." The slip into the Queen's Bath "had a drain, or small common sewer, running through it, which was raised like a bench and covered with a single board that it might serve as a seat! The baths were open to the winds and exposed to the gaze of idle spectators." These defects Wood proposed to remedy, but his scheme fell through. A piece of ground was required from the Duke of Kingston to make two fresh entrances to the Queen's Bath. His Grace consented to give the land, provided the Corporation would remove a shed lately built on the south side of the Abbey Churchyard. This they declined to do, fearing that if the shed were removed the Duke would be enabled to widen the street from the Abbey Green to the Churchyard. Wood maintains that his Grace could have made the road had he chosen, and left the Corporation in the enjoyment of their nuisance. When we contrast what the Baths are now, it must be a marvel that rank and fashion flocked to use them in the repelling condition above described. It shows how low the standard of social comfort and refinement was generally, and that to "rough it" was the common lot in the "good old times."

The state of the town within the walls was on a par with the Baths. There was "not a street, lane, alley, or thoroughfare whose sides are straight, but what has offensive gutters above ground, or water from the roofs discharged into it by spouts." The cleansing was left to be done by the rain, and the stables of the inns were so many or so situated "that there is scarce a line of buildings within the walls of the town but what looks into a stable yard!" Wood was never weary in declaiming against these things, and trying to spur the Corporation to do something to improve them. He was the first to urge the removal of the old Town Hall, which, with the addition of a "cage for prisoners" (especially for the exhibition of watchmen who were found in any house before 4 a.m.) blocked the centre of High Street. He wanted the Old Bridge widened, the narrowness of which was a nuisance, and lived to see it done; he advocated the demolition of the houses in the Abbey Churchyard, the area to be left open to Cheap Street and Stall Street, in order to bring more prominently into view the Mother Church and Pump Room. It will thus be seen that the great architect was a reformer born out of due season, one who was regarded as a troubler in a contented Israel.

If any self-satisfied individuals, in a vexed mood, exclaimed, " O that mine enemy would write a book ! " they found in due course their wish gratified. Wood did write a book, which he expanded to two volumes, and for a great part of it he has made himself a laughing stock. His mind was so saturated with Greek mythology and Druidical lore that he set himself the task of associating both with the fable of Bladud and his pigs, and the Prince's building operations in this locality ; while places in and around Bath were identified with Druidical rites and ceremonies, for the truth of all which he ingeniously argues, and strengthens by plans. He endeavours to show that Bath was the metropolitan seat of the British Druids, and that the stones at Stanton Drew form a perfect model of the Pythagorean system of the planetary world. He is inclined to believe that King Bladud's oracle stood in the Grove, where was a hollow tree from which mysterious utterances emanated, like those of Dodona. He indulges in pages upon pages of this fantastical nonsense, and with a faith in his lucubrations that enables him either to see no difficulties or to overcome them by far-fetched theories. A more astonishing example of the application of mythological erudition to support the chimerical is hardly to be found. One result of these wasted labours is that the Bodleian Library at Oxford possesses " descriptions of Stonehenge and Stanton Drew, by John Wood, of Bath, A.D. 1740 ; the former illustrated by five and the latter by four large plates." Apart from all this classical or pseudo-classical rubbish, Wood's book contains a mass of sterling information regarding the city, without the help of which many of its historic pages would be a blank. The author, who was a J.P. for the county of Somerset, died after a long and painful illness at his residence in Queen Square, on the 23rd May, 1754, in the fiftieth year of his age, " having by his professional labours secured a handsome competence for himself and family." His wife, a son, and two daughters survived him. Two years later the elder of the latter was married to Mr. Henry J. Coulthurst, of Melksham. She was described as " an agreeable young lady, with a fortune of £5,000." Mrs. Wood died at the house in Queen Square in April, 1766. The obituary notice adds, " by her death a considerable fortune devolves to her son, John Wood, Esq."

To Wood senior belongs the credit of having introduced the plan of devising a block of houses so as to give it the appearance of one harmonious whole, or " palace," as he terms it. His earliest and best example of this idea is the north side of Queen Square, the most admired, probably, of all his works. Its details are nothing like so elaborate as the Circus discloses ; but there is a breadth and dignity—a dignity without heaviness, and gained by simpler lines, that give to this section of the Square a charm that the Circus does not yield, beautiful as the latter is. Wood, be it remembered, made the design, and took the new departure above mentioned in building, when he was only in his twenty-fourth year. He also set the example of reserving open spaces in the midst of his buildings, which add so much to the beauty and salubrity of the city—an enlightened step which it is to be regretted has not been more generally followed here and elsewhere.

The great change wrought in the extent and appearance of Bath under the Woods was accompanied by a corresponding improvement in the furnishing of the houses. Wood, in his preface to the second edition of his " History " (published 1749), compares the interior of the dwellings then with their condition in 1727. As the new buildings advanced (he says) carpets were introduced to cover the floors, though laid with the finest clean deals or Dutch oak boards. The rooms were all wainscotted, and painted in a costly manner ; walnut tree chairs—some with leather, others with damask or worked bottoms—

supplied the place of such as were seated with cane and rushes; tables and chests of drawers were made of mahogany; handsome glasses were added to the dressing-tables; nor did the proper chimneys or peers of any of the rooms long remain without well-framed mirrors of no inconsiderable size; the linen for the table and bed, as well as the window curtains, grew better and better, till it became suitable even for people of the highest rank." With all this increase of goodness (he adds) lodgings received no advance during the seasons, which, however, were almost every year lengthened. In conclusion he remarks, "To make a just comparison between the publick accommodation of Bath at this time (1749) and one-and-twenty years back, the best chambers for gentlemen were then just what the garrets for servants now are." These details are interesting, as showing in a vivid manner the domestic economy of the city in the first half of the last century, and as furnishing materials for drawing a comparison with what it is now. The progress in the interval has been marvellous, and in no particular more than in the all-important factor, sanitation, of which Wood makes no mention. He gives the number of houses as 1362, and estimates the total population at 9,000.

WOOD THE YOUNGER.

THE son, of whom we have now to speak, had the advantage of being trained under his father. He possessed much of his sire's energy, and a full measure of his talent. His first essay was to get the Circus built, which was begun in 1754, but was not completed for another fifteen years, the sites being slowly taken by the public. In the meantime Wood had planned and built Brock Street (where he resided after leaving Gay Street), Margaret's Chapel, and Margaret's Buildings, besides completing his design for the Royal Crescent, and issuing a prospect in November, 1764, for building Assembly Rooms at the north-west corner of Queen Square. It was to be "a large building, consisting of a tavern, a coffee room, and a complete set of assembly rooms," the cost not to exceed £16,600, divided into 120 shares, on the tontine principle. It is strange to find a "tavern" occupying a primary position in such a scheme, but it was accorded the honour out of deference to the convivial weakness of the age; albeit some objected that it would be "an improper appendage to a sett of public rooms." Wood explained that the tavern could be applied to another use if the majority of the subscribers desired it. Promises for the amount of capital required were secured by the end of the following April; but the undertaking remained in abeyance, and Wood revised his plans, which now provided "a complete set of rooms for the accommodation of the nobility and gentry resorting here." They are to be built "on the east side of the Circus, in Mr. Holdstock's garden, the quantity of ground, including the court for chairs and the approaches for coaches, being upwards of an acre and a half. The capital to be raised by 70 shares at £300 each." The first stone was laid on the 24th May, 1769, by Wood himself, and the building was ready for opening in October, 1771, the total cost having been about £20,000. The rooms are a model of completeness and compactness, the whole space having been admirably utilised to meet the convenience of the patrons, while the principal rooms are finely proportioned, and their decorations chaste. They are splendid testimony to Wood's common sense in making internal arrangements, just as the façade of the Royal Crescent exhibits his consummate skill in design externally. The latter was begun in May, 1767, and completed in about eight years.

In the former year Wood competed to build a new Guildhall and markets, but neither of the three or four plans sent in met with approval. He was afterwards asked to submit a new design, which likewise was not deemed satisfactory in point of cost. He, however, received a solatium of thirty guineas, and a year or two later was instructed by the Corporation to erect a new suite of Baths in place of the old Hot Bath, in Hot Bath Street. To increase the area several ramshackled tenements were removed, and this enabled Wood to provide a very useful and in some respects ornamental bathing establishment, which has been extended and improved in modern times. He also nearly doubled the yield of the spring; the water from it percolating away through the loose rubble that covered it, he made a large stone cylinder, eight feet in diameter and fifteen feet deep, to enclose the source, which increased the supply from 80 tons in the twenty-four hours to 140 tons. For his services he was voted 100 guineas. Among other buildings which he designed may be mentioned the York House, Alfred Street, and Fountain Buildings. He also rebuilt Woolley Church; but the style is not to be commended. Wood's life soon came to a close. He died at his house at Batheaston on the 16th June, 1781, and, like his father, was buried at Swainswick; but neither tomb nor tablet marks the spot. The absence of a memorial of this kind made it more imperative that the houses occupied by them in Queen Square and Gay Street should have identifying tablets affixed to them. This was done in the autumn of 1900, under the scheme carried out by the Corporation of making known all houses of historic interest. Wood left a widow and a daughter, Mary, who, in the following November, married James Tompkinson, of Dorfold, Cheshire.

The final glimpse we get of the family is a sad one. In 1807 Mrs. Elizabeth Wood, the widow just mentioned, was living at Richmond, Surrey, in great poverty. What became of the wealth her husband and his father undoubtedly accumulated it is impossible to tell. Her condition shows that it had been alienated or dissipated. She was then in her eightieth year. In her distress she appealed to the Bath Corporation, stating that she had little or nothing to live upon, and hoped the recollection of her husband's and his father's services to the city would induce them to give her some assistance. Without hesitation the Council ordered that a sum of £20 should be paid to her yearly during the term of her natural life, the first remittance to be made immediately. No doubt this generosity helped to render comfortable the closing years of the last of the Wood family.

It has been finely said that the wings of life are plumed with the feathers of death. Largely plumed they were in the case of both the Woods, seeing how short, comparatively, was the career of each. Brief as it was, their names and services will not be forgotten. The buildings raised under their auspices still command admiration, despite the fluctuations of taste and the changes progress evolves. They bear the stamp of the rare genius of two men who re-created Bath in the eighteenth century, and "time's effacing fingers" are not likely soon to obliterate it.

THOMAS BALDWIN

THE improvements in the city and the additions made to its public buildings already recorded were greatly augmented by the labours of Thomas Baldwin, whose genius was little inferior to that of the Woods, of whom he was a faithful disciple, so much so that their works can scarcely be distinguished the one from the other. It was fortunate that Baldwin had imbibed the same taste for the renaissance, inasmuch as it gave uniformity to the architectural features of the city, yet with sufficient variation in details to exclude monotony. In his career he was favoured by circumstances that gave scope for the exercise of his skill. Just when his reputation was established, the Corporation, stimulated by the handsome appearance the upper town presented, determined to make the old town less incongruous and more convenient; to remove obstructions, widen streets, and remodel the Baths and Pump Room. An Act of Parliament was obtained authorising these works, the cost of which was estimated at £80,000. Simultaneously with this change in the attitude of the authorities, the Bathwick side of the river, chiefly open meadows, was offered for building by the Pulteney family, who erected the bridge bearing the name to connect the estate with the city. The celebrated Adams designed the bridge; but Baldwin was commissioned to prepare the plans for the intended streets—a task he executed, as all can see, with marked ability.

Born in 1750, his talents, like the elder Wood's, were early developed, as in 1774 he considered himself competent to build the Guildhall and markets, a prize he secured. The history of the undertaking unfolds a curious chapter in municipal history. The old Town Hall built by Inigo Jones in 1629 stood, as stated, in the centre of High Street, and was flanked on either side by an irregular line of houses. One end of this contorted parallelogram was enclosed by the north gate, while at the other end stood the tenements which then and for long afterwards encrusted the front of the Abbey. As early as 1760 the Hall was so dilapidated as to be considered unsafe, and the Corporation turned its attention to re-building it, being encouraged by the offer of £500 towards the expenses from Ralph Allen. Six years, however, elapsed before the foundation stone was laid, and then the work was stopped owing to the difficulty of getting possession of some houses which had to be cleared away to give the space required. At length these obstacles were removed, and the Corporation, thinking they had an architectural genius in their body, commissioned him to undertake the work. Thomas Warr Atwood was his name; plumbing and painting his profession, to which he added a general supervision of the corporate property. He had, however, built a house for the Pulteneys on the Bathwick side of the river (afterwards the gaol), and this was considered to justify the entrusting him with the civic buildings contemplated. His plans were approved, as well as the estimated cost, £6,500, five per cent. on which being his remuneration. This, it was said, he sought to augment by privately placing the contracts where he could secure a commission of equal amount. When he did invite tenders, tradesmen complained that the interval allowed was so short that they had not time to prepare estimates, even if they understood the specifications, which they did not.

John Palmer, who built the Theatre in Orchard Street for his namesake, John Palmer (the lessee), insinuated that he had discovered "gross errors in the accounts of a certain person to whose management the Corporation affairs had been entrusted to a degree little short of infatuation." He intimated his readiness to produce plans for a Hall and market, which should be more convenient and cost £2,000 less. The Council thereupon passed a resolution vindicating Atwood from these "scandalous and malicious insinuations," and directing that his plans should be carried into execution forthwith. Palmer then made a more taking proposal. He would erect the buildings for nothing if the authorities would grant him a 99 years' lease of the land on which certain shops were to be placed. The public mind was still further inflamed by this apparently generous offer. The Freemen rebuked "their trustees" (the Corporation) for squandering the public money, and threatened to appeal to the House of Commons! Others accused the Building Committee of "employing none but their own creatures," and "urged them to put a stop to their disgraceful plan."

In the midst of this wordy warfare a sad event occurred. Atwood, while superintending the pulling down of one of the old houses, was crushed by a falling floor, and received such severe injuries that he died. His death hushed the controversy, and during the lull it was announced that a plan would be submitted that would include the merits of both those in dispute. The fresh draft was by Baldwin, who appears before to have posed as an "amicus curiæ" between the opponents. It was adopted by the Corporation, but, strange to say, the minutes of that body are silent on the subject. The resolution in favour of Atwood's plan was not rescinded, neither was the new departure recorded. Either it was deemed unnecessary or the members felt chary in embalming in their archives a defeat which they had sustained through popular clamour, although it may be assumed that it was highly beneficial, as it gave to the city a Guildhall and markets of which it had every reason to be proud. The exterior of the former is harmonious and dignified, while the interior is admirably planned: it has a noble banqueting-room and provided all the offices necessary for the convenient transaction of civic business until the growth of the latter made extension necessary. How far Baldwin was indebted to the earlier designs cannot be answered; possibly not much. Palmer had been bred a carpenter, and though he raised himself to be an architect he was without the inspiration of originality. Atwood's qualification may be similarly gauged. The point invested with more doubt is whether Wood's plan of 1766 may not have supplied important hints to the subsequent competitors. However, engravings published at the time gave Baldwin the credit of being the architect, and his claim was not challenged. That the Corporation were satisfied is shown by the fact that he was appointed City Surveyor in 1780, and City Architect in 1791, when the general renovation of the old town was in progress.

Prior to this time some important improvements had been made. Thus the north and south gates were removed, as well as the walls and some houses adjoining them, in 1755; the avenues of the west gate were also widened, and the Old Bridge was doubled in breadth. The greatest evil still remained—viz., the want of proper communication between the upper and lower town. The completion of Milsom Street and Old Bond Street, which brought the northern thoroughfares to the confines of the old town, accentuated the inconvenience. Milsom Street, broad and handsome, was, with the Octagon Chapel, built by Lightholder. The east side, with its semi-circular columnar centre, supported by pediments on either hand, is very striking, and was more so, of course, before the medley of projecting shop fronts made its appearance. These additions on both sides narrowed the street and damaged the architectural effect. From it no direct way into the heart of the city existed. There was no New Bond Street, which was known as Frog Lane, and no Union Street, which was then the site of the stable yard of the Bear Inn—an inn facing Stall Street and connecting Cheap Street and Westgate Street. The former was so narrow that two vehicles could not pass each other, and the latter was not much better. The old White Hart Inn (now the Grand Pump Room Hotel) encroached considerably on Stall Street; Bath Street was a mean thoroughfare; the Pump Room, though it had been improved, was still inadequate, and reached as it was by a flight of steps, was found very trying to invalids. A miserable avenue, obstructed by houses and stables, as well as by the West Gate, led from Queen Square to the lower part of the city. Thus the Bear Inn premises provided a short cut from the above Square to the Baths, and explains the use made of it, as described by Smollett in "Humphrey Clinker." "The communication with the Baths is (he says) through the yard of an inn, where the poor, trembling valetudinarian is carried in a (Sedan) chair between the heels of a double row of horses coming under the currying of grooms and postillions, over and above the hazard of being obstructed or overturned by the coaches which are continually making their exit or their entrance."

It was to make a clean sweep of all these evils that the Act of Parliament before mentioned was obtained. Under its powers the Bear Inn and its appurtenances (purchased for £13,750) were removed; Union Street built, Cheap Street and Westgate Street widened, the White Hart set back several feet, Bath Street (with its Doric colonnade) raised, and the Cross Bath at the end—a charming edifice—reconstructed, and Hetling Pump Room built. The whole of this work fell under the guiding spirit of Baldwin; while to Palmer was entrusted the making of a new thoroughfare from Trim Street to the South Parade, for which preparation had been made by the demolition of the West Gate in 1776. As City Architect, Baldwin had to supply the plans required for the Grand Pump Room in Abbey Yard, with the piazza and new entrance in Stall Street. The latter features were no doubt his handiwork, and tasteful they are; but with regard to the Pump Room the same or even greater difficulty arises in apportioning his share of the credit as in the case of the Guildhall. Before he was appointed City Architect, the Corporation had commissioned Reveley (a pupil of Sir Wm. Chambers), who built Camden Crescent, to provide plans for the Pump Room and the Baths adjoining. He prepared designs of "great beauty and elegance," and a beginning was made with the work, a considerable portion of the elevation on the west side being finished. Progress was then arrested, and Baldwin was instructed to make a fresh plan, probably one less costly, which he did in due course. Not long after (in the autumn of 1791) serious differences arose between him and the Corporation on, it is assumed, financial matters. He was ordered to deliver up all books, papers, etc., in his custody, or in default to have a bill in Chancery filed against him. He declined to comply, and in July, 1792, the Town Clerk was directed to commence legal proceedings for the recovery of the documents retained. Baldwin appears to have then yielded, and made the surrender demanded. The Pump Room Committee investigated matters, with the assistance of Mr. Palmer, and reported that £2,000 was still required to finish the Pump Room.

Acting under instructions, Palmer provided fresh plans and elevations, which were adopted, and the Room was ready in 1799. How much of the building is to be ascribed to Reveley (who was paid £28 for his plans), and how much to Baldwin and Palmer is a problem that cannot now be solved. Baldwin lived for some time in Pulteney Street, and died in 1820.

There is one fact connected with the building operations in Bath in the last century which, already glanced at, merits particular mention. It is the facility with which men engaged in ordinary handicraft occupations developed into architects. The evolution of these untrained professors was probably due to the rapid growth of the city, and the stimulus thus afforded to any building talent latent. Thomas Greenway, who, in 1707, built the Cold Bath House in Claverton Street, and later the Garrick's Head, was a free-stone mason; William Killigrew, who designed the first Blue Coat School, was a joiner; John Harvey, who re-built St. Michael's Church, was a stone carver; John Palmer, who built two or three churches and other public buildings, was a carpenter; Thos. Warr Atwood, who furnished the plans for the prison in Grove Street (later the Police Barracks), was a plumber and painter. From these examples it might be inferred that architects, like poets, are born, not made. Wood the Elder had a contempt for such amateurs, and insists, for their benefit, that "an architect should be well grounded in the theory and practice of geometry," if he would avoid mistakes. Still, some of their works are not to be despised, as Rosewell House (Kingsmead Square), Weymouth House, and three or four other dwellings near St. James's Church and in Westgate Street attest. They are interesting examples of domestic architecture, dating from different periods of the century.

Another artisan who is represented as acquiring the position of an architect is Richard Jones, but his credentials for admission into the hierarchy of the untutored are very vague. He was a mason by trade, and for many years was a foreman in Allen's works. Just before his death he wrote a confused autobiography, in which he asserts that he built Prior Park, with the exception of the first story of the mansion, the Palladian Bridge, and other important works for Allen and others. No one apparently ever heard of his achievements until these reminiscences saw the light, and the probability is that Jones, in his dotage, was under the delusion that he was the supreme artificer of works of which he only had the superintendence in execution. The late Mr. Peach, unfortunately, accepted as solid facts these idle vapourings of senile decay, and in one or two of his works he gives Jones full credit for all he claims to have done. The best evidence of Jones's status is to be found in Ralph Allen's will. He is there classed among the menial servants who were to receive one year's wages, which, in his case, amounted to £45. Now, if he possessed the ability he vaunts, Allen, who was a generous employer, would most assuredly have given him more than 17s. 6d. per week, or if Allen had behaved meanly, Jones would with equal certainty have carried his talents to a better market. Moreover, just after his master's death, he was engaged, he tells us, in building some pigstyes by the order of Mrs. Warburton. A drop from alleged Palladian Bridge designing to actual pigsty building should, with his legacy, have been fatal in the eyes of anyone to his professional pretensions. These having, however, obtained recognition in the works of a local authority, are likely to be esteemed well founded. All that can be done to correct the error is to enter a caveat when occasion offers against the apocryphal performances of a self-deluded old man.

The building activity already noted was operating over a much wider area. On all sides, with the exception of the Southern, new streets had sprung up. The ancient Ambry of the monastery, lying between the town walls and the river, was covered with houses. In the Kingsmead, New King Street, Greenpark Buildings, and adjacent streets, with Norfolk Crescent, had been erected. Higher up, Marlborough Buildings, St. James's Square, Somerset Place, and Lansdown Crescent had taken the place of fields and gardens. Lower down, Portland Place, Burlington Street, Rivers Street, Belvedere, and Camden Crescent; further to the east, Kensington, Grosvenor Place (with its hotel and gardens), and Beaufort Buildings; while a new town had sprung up in Bathwick under the hand of Baldwin, who had the satisfaction of knowing that Pulteney Street was pronounced one of the finest streets in Europe. Rapid as was the growth of the city, its expansion bid fair to continue; but the war with France, which began in 1793, soon checked enterprise. Buildings were left half finished, or with only their external walls, and so remained until after the return of peace, some of the builders having been ruined in the interval. As it was, the city at the close of the century wore an entirely modern aspect, nearly every vestige of its ancient and quaint structures, with the exception of the Abbey, Hetling House, Bellott's Hospital, and the Talbot Inn, having disappeared. The changes made gave it the palatial character it still wears and secured for Bath the proud title of the "Queen of the West." The population had also increased to 34,160—13,790 males and 20,370 females—and the houses to about 6,000.

CHAPTER II.

FASHION AND THE RULE OF THE M.C.

BEAU NASH.

EFORE the commencement of the eighteenth century the desirability of making Bath a city of pleasure as well as a health resort does not appear to have been suggested or entertained. The Corporation, as the custodians of the Baths under the charter of Queen Elizabeth, naturally considered that their responsibility in regard to visitors began and ended with the keeping of the bathing establishment in a usable condition, according to their lights. Their economical bias was, however, strong; bare utility was the standard they kept in view, and as improvements cost money they were not regarded with favour, or were shirked on the slightest pretence. A stimulus was needed to arouse them to greater effort, and this was supplied by the evolution of the M.C., especially in the person of Beau Nash, who practically created the office. The best sketch of Nash is to be found in a scarce pamphlet entitled "Characters at the Hot Well, Bristol, in September, and at Bath, in October, 1723." For the benefit of posterity the writer records that "Mr. Nash is a man about five feet eight inches high, of a diameter exactly proportioned to your height, that gives you the finest shape; of a black-brown complexion that gives a strength to your looks suited to the elastic force of your nervous fibres and muscles. You have strength and agility to recommend you to your own sex, and great comeliness of person to keep you from being disagreeable to the other. You have heightened a great degree of natural good temper by cultivating the greatest politeness, which, improved with your natural good wit, makes your conversation as a private person as entertaining and as delightful as your authority as a governor is respectful. With these happy accomplishments, with the fine taste you discover in whatever habit you please to appear, and great gracefulness with which you dance our country dances, it will be no great wonder that you support your empire when once you obtained it. I don't mention your dexterity in French dances, because you don't affect dancing them; in which, I think, you show your judgment, though, no doubt, you might as well excel in a Minuet or Riggadoon as in 'Bartholomy Fair' or 'Thomas I Cannot.' To conclude, I write this about the eighteenth year of your reign, and the eight-and-fortieth of your age, from my lodgings in the Grove at Bath." The writer further bears testimony to the Beau's success in maintaining "decency and order in such a mixed society of persons of all ranks, orders, and sexes, as compose the general resort to the waters, notwithstanding two such great 'Makebates' stand in your way, I mean play and women." In conclusion, the author tells us that he has just seen Nash go into Hayes's rooms, in his usual "very agreeable oddness of appearance—black wig, white hat, scarlet countenance, and brown beaver habit."

Such was the remarkable individual whose name is indissolubly connected with the prosperity of Bath. Not that he confined his attentions to one place. "Though with respect to the greater number of your subjects (says the pamphleteer) you are King in Bath, you may be styled Prince at the Hotwell, Duke at Tonbridge, Earl at Scarborough, not to mention your Lordships of Buxton, and your own kindred's (Nash was a Welshman) famous place of resort, St. Winifride's Well." However agreeable he may have been in person and demeanour, Nash was no Puritan in his conduct, as may have been gathered from the authority above quoted. A libertine and a gambler, his vices were nevertheless overlooked in an age which regarded them as the mere peccadillos of the fine gentleman. Yet, because his authority was absolute, he has been censured for not discouraging the profligacy rampant under his sway. The effort would have been futile, even if Nash had thought it his duty or had had any inclination to make it, which he had not. His sole object was the correction of manners, not morals; public decorum, not private virtue, he sought to promote. This limitation of his rule was acquiesced in; to have gone beyond it would have kindled resentment. In fact, any attempt to stem the tide of immorality, which flowed towards the city, would have been no more efficacious than was Dame Partington's broom to check the Atlantic breakers. Nor should it be forgotten that Nash's power had no foundation beyond prescription. He was practically a usurper, without any valid title. Had he interfered with the private vices of his subjects he would have been denounced as a meddlesome hypocrite, have been speedily deposed, or had a rival set up in opposition to him. His supremacy, therefore, depended on the toleration of evils which he himself did not regard with the disgust an age of more refinement views them. In fairness he must be judged by the ethical standard of his time and not by that of subsequent generations.

The real question is whether he succeeded in effecting the reforms which he aimed to accomplish. The answer must be in the affirmative. His task was by no means an easy one. Instead of the select few who came here prior to the eighteenth century, there was, as already mentioned, a rush of all ranks and degrees, the bulk of whom came to enjoy themselves. There was no distinction of classes; no exclusiveness. All met on an equality. The ability to pay for the pleasures provided was the only test for admission. With the boorishness common among the men, and the hoydenish ways of the fair, to say nothing of the roystering Mohocks bent upon indulging their rowdy instincts, nothing but an iron hand, with a silken glove, could possibly have prevented degradation and disorder being rife among such a motley and self-willed society. Under these conditions Bath would have been shunned except by the vilest, and its history, at least for some time, would have been a sorry one. Nash, at the outset of his career (1704), saw the danger and knew that it could only be averted, and the place made attractive, by decency and politeness being observed by visitors whatever their social status. Thus the rustic gentry, accustomed to follow the bent of their own inclinations, and with no idea of manners beyond what their home surroundings supplied, found themselves, when they came to Bath, the subjects of strict discipline, both as to their behaviour and dress. Respect for the authority of the M.C. was exacted in the most trifling details, under a penalty of exclusion from the assemblies in the event of any infringement. The direct benefits the company thus enjoyed from these wholesome regulations were, it must be remembered, supplemented by the indirect advantages society at large derived from the educational influence of Nash's policy; it was a leaven working for the refinement of manners generally. If, too, such discipline was necessary for the comfort of the pleasure-seeker, how much more so was it needed for the protection of the health-seeker—for the invalids who sought in the baths relief for their bodily ailments? In their case the reign of law was of primary importance, to secure them from the shocks to which they would have been exposed from unbridled licentiousness. Nash realised what a rich mine lay in the Bath waters, and he lost no time in trying to make them popular, so that their latent wealth should contribute to the material well being of the community, not without an eye, perhaps, to his own profit and renown.

How ready he was to avail himself of every opportunity to extend his power is shown by his artifice to allay the apprehension caused by the declaration of the "greatest physician of his age," that he would ruin Bath by casting (as he said), a toad into the springs. Nash had read Dr. Mead's treatise on the power of music in the poison of the tarantula. The band then consisted of five very indifferent performers, who played in the Grove, and at the Cross Bath. He imported seven professional players from London, stating that "he would fiddle the amphibious creature out of the hot waters; and, by the power of harmony, charm every one on whom the toad should spit his poison into such a dance as should drive out the venom, and turn languishment itself into gaiety." The physician's machinations were, of course, defeated, and a good band was substituted for an inferior one. A necessary change was thus covertly effected, and the dread of an impending disaster removed from the minds of the credulous public.

At this time there was no Pump Room and no Assembly Room. A common booth was placed on the Bowling Green whither the company resorted to play cards and to drink their "tay" and chocolate.

Here also they were wont to dance, exposed to the weather and the gaze of gaping crowds until, at the instigation of the Duke of Beaufort and with the consent of the Corporation, these Terpsichorean amusements were, at the close of the seventeenth century, allowed to be held in the Guildhall. The urgent need of having some rendezvous for visitors was so obvious that Nash had very little difficulty in inducing Mr. Thomas Harrison to build a room for their accommodation on the east side of the Grove, with access to the ornamental grounds below, known as "Harrison's Walks."

The Corporation, after the visit of Queen Anne in 1703, also began to think that some shelter would not be a bad thing for the bathers and water drinkers, exposed as they were to the rain and cold, unless they returned at once to their lodgings. The work was not an easy one to accomplish, as the authorities did not own a foot of land outside the walls surrounding the baths. They met the difficulty by purchasing two or three houses on the south side of the Churchyard, and in the space thus gained the first Pump Room was erected, £100 being given towards the cost by Dr. Bettenson. The architect was John Harvey, whose father, according to the Corporation minutes, was paid for carving the capitals of the building. The room was completed in 1706, and was opened by a public procession and a musical fête, the following composition being included in the programme:

"Great Bladud, born a Sov'reign Prince,
But from the Court was banish'd thence
    His dire disease to shun;
The Muses do his fame record,
That when the Bath his health restor'd,
    Great Bladud did return.

This glorious Prince of royal race,
The founder of this happy place,
    Where beauty holds her reign;
To Bladud's mem'ry let us join,
And crown the glass from springs divine,
    His glory to maintain.

Let joy in every face be shown,
And fame his restoration crown,
    While music sounds his praise;

His praise, ye Muses, sing above;
Let beauty wait on Bladud's love,
    And fame his glory raise.

Though long his languish did endure,
The Bath did lasting health procure,
    And fate no more did frown;
For smiling Heaven did invite
Great Bladud to enjoy his right,
    And wear th' imperial crown.

May all a fond ambition shun,
By which e'en Bladud was undone,
    As ancient stories tell;
Who try'd with artful wings to fly,
But towering on the regions high,
    He down expiring fell."

To modern readers these lines savour of the burlesque; but it should be remembered that when they were written and sung, the Bladud myth—the pigs, the cure of the prince, his founding Bath, and his Icarus-like fate—was deemed, at least locally, veritable history, to doubt which was unbecoming and unpardonable presumption.

The room was placed in charge of a pumper, who paid the Corporation an annual sum for the post, he reimbursing himself by the gifts received from visitors. Mr. Nash's band, as it was called, was, on the request of the doctors, transferred to the new room, the salaries of the performers being paid out of the subscriptions received for the balls. A pavement was likewise made in the Grove, with large flat stones, for the company to promenade upon. Nor did solicitude for the health and comfort of visitors end here. Nearly eighteen hundred pounds were raised by public subscription to make a good road to Lansdown, that the "invalids might conveniently ascend the hill to take the benefit of the air"; they were also exempted from all manner of toll as often as they went out of the city for "air and recreation."

With these improvements, combined with Nash's skilful administration, the number of visitors annually increased. To meet their requirements Harrison, in 1720, added a large ball room to his house; it was 61ft. 6in. long, 29ft. broad, and 28ft. high. A second Assembly Room was built in 1729 by Mr. Thayer from Wood's design, at the east end of what is now York Street, the ball room of which was a double cube of 30ft. Neither of these places was, however, convenient or structurally attractive. Subsequently, Wood was anxious to build an Assembly Room on a scale commensurate with the requirements of the visitors, and in harmony with the new edifices he had designed. Nash was equally convinced of the necessity and urged him to give practical effect to the scheme. A "wild proposal" to enlarge Simpson's rooms stayed Wood's hand, and his death shortly after left the project to be carried into execution by his son, who, as we have seen, did so in a masterly manner.

In the meantime the Pump Room was found to be inadequate for the company, and it was pulled down, another being erected in its place in 1732, it was in length 43ft. 6in. by 26ft. It was enlarged two years later and a gallery built for music; a marble cistern was also substituted for the copper one formerly in use.

All the while Nash was moulding the promiscuous and growing body of visitors into habits of politeness and self-restraint. His first code of rules for regulating manners, eleven in number, was hung up in the first Pump Room. They read like persiflage—half in jest, half in earnest; but witty they were deemed at the time, and this made them the more acceptable. Later he added oral edicts to correct what he regarded as abuses or irregularities. After a struggle he succeeded in compelling the discontinuance of the practice of wearing swords. He was aided in this reform by the occurrence of a duel with swords between two gamesters (Clarke and Taylor). They fought by torchlight in the Grove. Taylor was run through the body, but lived seven years after. Clarke professed to be a Quaker, and died in poverty in the course of eighteen years. Nash would have no more fighting, and whenever he heard of a challenge given or accepted he instantly had both parties arrested. He prohibited country squires from coming to the balls in top boots and the ladies in riding hoods. If he espied a booted gentleman he made up to him and, with a polite bow, blandly informed him that he had forgotten his horse. He was inflexible in exacting respect for his rules from all ranks. There is the well-known instance of his removing from the Duchess of Queensberry a costly white apron and tossing it aside with the remark "that none but Abigails appeared in white aprons." Her Grace, instead of resenting it as a rudeness, was abashed at her indiscretion, and begged the Beau's pardon. Surprise at this submissiveness is not lessened when it is remembered who the Duchess was. She was the charming and wayward Catherine Hyde, grand-daughter of the Earl of Clarendon—the Kitty, whose first appearance at Drury Lane Theatre as a triumphant beauty of eighteen, Prior had celebrated in some of his brightest and airiest verses, and whose picture (says Mr. Austin Dobson) as a milkmaid of quality, by Charles Jervas at a later date, may be seen at the National Portrait Gallery. Nor did she mind the frowns of a real monarch, as was shown by her canvassing for subscriptions to the "Beggar's Opera" in the presence of George II., to whom the supposed political allusions in the work were distasteful, conduct for which she was banished the Court. In like manner when the Princess Amelia wanted one dance more after Nash had given the signal to withdraw he refused, telling her that the laws of Bath resembled those of Lycurgus, and would admit of no alteration.

These were master strokes of policy. Nash saw the opportunity afforded him of blazoning abroad his authority, and he readily seized it. When grandees were snubbed the commonalty viewed with greater awe than ever the titular King of Bath, who could exact obedience to his orders from those of exalted rank. No doubt there was a considerable leaven among the pleasure seekers who had the good sense to see that without authority there could be no law or order, and who, on that account, were grateful to the uncrowned monarch for his rigid legislation and administration. Lacking such support he could never have obtained the autocratic power he wielded in social affairs. That he was not always polite is shown by the incident of his tossing into the King's bath a gentleman whom he overheard admiring his wife's charms as she was bathing. It was a bad example for a law giver, but was due, perhaps, to the recrudescence of an impulse from his wild oat days.

The object of Nash's labours was to make the etiquette of his little realm as formal and stately as that of the Court of St. James or Versailles. Each ball, which began at six, opened with a minuet danced by two persons of the highest distinction present. When the minuet concluded, the lady retired to her seat and the M.C. brought a new partner to the gentleman. The like ceremony was observed by every succeeding couple, each gentleman having to dance with two ladies till the minuets were over, when the country dances ensued, ladies of quality, according to their rank, standing up first. At nine o'clock there was a short interval for rest, and the gentlemen helped their partners to tea. Dancing was then resumed until the clock struck eleven, when the M.C. ordered the music to desist by holding up his finger. Moreover, the day's routine, as well as the etiquette of the balls, was governed by Nash's arrangements, which had the countenance of the medical profession. The custom was for ladies and gentlemen to meet in the Pump Room at eight in the morning to drink the water or converse; the band played until ten, the hour for breakfast, which was obtained either at lodgings or at coffee houses, one or two of which were for the exclusive use of females. Concert breakfasts were sometimes given at the Assembly Rooms, "persons of rank and fortune who can perform being admitted into the orchestra." The Abbey was open for morning prayer at eleven, and after midday every one took exercise, some promenading on the parades, others going for a ride in Hyde Park (or the Common) or a drive to Claverton or Lansdown; more adjourning to the Kingsmead, then a verdant expanse, with the pellucid Avon on its margin, where pleasant walks could be enjoyed, and at several cake-houses refreshments obtained, including fruit, "lullibubs, and sumes liquors."*

Three o'clock was the common hour for dinner, which left the evening free for another visit to the Pump Room and for the balls and concerts; as these were not allowed to continue after eleven, in order to avoid undue fatigue, everybody was, or could be if so disposed, at home before midnight.

There is, however, a dark side to this somewhat idyllic picture. A glance inside Simpson's or Wiltshire's rooms at nocturnal hours would have disclosed scenes of sordid knavery and infatuation. The rattling of the dice-box, the shuffling of cards, the rippling of balls at the E. O. tables, the jingling of coin, the clinking of cups and glasses were to be heard, while in the dim light shed by flickering candelabra could be seen men and women at the different tables engaged deeply in play, haggard in mien for the most part from the terrible tension betwixt hope and fear they were undergoing; there a face flushed with a transport of delight at an unexpected turn of the game; there a ruined gamester, with a bewildered expression of despair, tottering from the room. Suppressed excitement is all-prevailing, save where sharpers sit cool and collected, or indulging in mock sympathy with their victims. And Nash? he, too, is there displaying his customary "insouciance," trying, it may be, to check the desperate stakes laid by some youthful plunger. Hypocritical zeal it must be, for is he not leagued with the croupiers, in whose rich harvest he expects to share? A haunt of pleasure this! What a mockery that it should be filled with a company racked with suspense, and over some of whom dire disaster broods!

Such was the daily programme which visitors were expected, and did for the most part, follow. Nash's supervision and authority extended even beyond amusements. If a fire broke out it was Mr. Nash who took an active part in extinguishing it; if a subscription were needed for the sufferers, he started it; if a highway robbery occurred in the neighbourhood it was Mr. Nash who despatched men to capture the footpads; if a festive celebration were required he arranged the details. When, too, the weavers from Twerton came begging in a body to Nash he gave them a bountiful dinner and a week's subsistence on going away. When the miners from the coal districts, impelled by want, came to the city yoked to a waggon laden with coals, which they sagaciously presented to the M.C., he opened a fund for their relief and headed it with a donation of ten

---

* For a Bath bun or buttered roll 4d. was charged, a "dish" of chocolate 6d., a "dish" of tea 3d., a cup of coffee 3d.

guineas. He collected more money than anyone else for the building of the Mineral Water Hospital, and for several years he was its appointed treasurer. What wonder that he was popular among the masses when all saw such evidence of his kindness and generosity! The Corporation exhibited its gratitude by placing a full-length marble statue of the Beau, with a plan of the Hospital in his hand, in the new Pump Room on its completion in 1751. It can still be seen in the present Pump Room, although disguised by sundry coats of paint, with which the effigy has been favoured. A portrait of him, life size, was hung in Simpson's Ball-room, and another in Wiltshire's Ball-room, the latter between the busts of Newton and Pope, which gave rise to the severe but witty lines of Jane Brereton:

> Immortal Newton never spoke
>   More truth than here you'll find;
> Nor Pope himself e'er pen'd a joke
>   Severer on mankind.
>
> This picture placed these busts between
>   Gives satire its full strength;
> Wisdom and Wit are little seen,
>   But Folly at full length.

At first there was only one Bath season, and that in the summer months, the roads then being in a better condition for travel. With the improvement of the highways, and the growing number of visitors, it was found possible and expedient to have two— one in the spring (March to June) and another in the autumn (October to Christmas); the duration of each, however, was liable to change. When the company had left, the streets were comparatively deserted, and dullness ruled. In the early days of Nash it was a common expression that a culverine, charged with grape shot, might be fired from the Bear Inn down Stall-street at noon without killing anything but a pig or a turnspit dog. The canine breed just mentioned was very largely represented, every lodging-house having one or more to help roast the meat. The animal was shut in a wheel, which it had to turn, like a squirrel in a cage, and thus supplied the motive power to keep the joint in motion. A Spencer of the Marlborough ilk, who was known to be a man of humour, once employed the chairmen to collect all the turnspits in the city together about one o'clock, and shut them up till four in the afternoon, causing by their enforced absence from duty no small confusion both in the kitchen and dining-room. Bishop Warburton must have been in an equally waggish mood if he really told the story, which Thicknesse placed to his account. He declared that he was at the Abbey Church one Sunday when a certain chapter in Ezekiel was read, in which the word "wheel" is mentioned several times. A number of turnspits, which had followed the cooks to church, manifested symptoms of alarm when the reader the first time uttered the hated monosyllable, but upon its being repeated twice more they all clapped their tails between their legs and ran out of the church. Not to be outdone in this drollery, Thicknesse adds that the turnspits, like their betters, held meetings in a part of the city, when they passed wholesome resolutions for their mutual benefit! The introduction of the smokejack deprived these short-legged quadrupeds of their occupation, as well as released them from the scullion's tyranny.

Nash cared nothing for the turnspits, their numbers or their treatment. His time and attention were occupied with the important duties devolving upon him as "arbiter elegantiarum." The pleasure the office gave him was intensified whenever Royalty came within his domain. His own dignity was enhanced, a nimbus of glory encircled his head, and the debt of gratitude due to him from the Corporation and citizens was increased, for was it not through his instrumentality that these scions of the purple patronised Bath? At least he was always ready to yield to the soft impeachment when it was insinuated or directly made. His first, though unofficial, experience of a Royal visit was when Queen

Anne and her consort, Prince George of Denmark, came here in 1702. He was probably among the crowds that flocked to the city from various parts of the country to see the loyal festivities extemporised for the occasion. If so he must have been an eye-witness of the pageantry when the Queen and the Prince were met on the confines of Somersetshire by one hundred young men of the city, uniformly clad and adorned, and two hundred young women gaily attired as Amazons, and by them were escorted to the western gate of the city, where they were received by the Corporation, and presented with an address, their apartments being over the gate.

The first Royal visitor that Nash had the honour of welcoming was the Princess Amelia, second daughter of George II., in 1728. She was met by one hundred young men uniformly equipped, received by the Corporation and Mr. Nash with due formality, and lodged also over the Westgate. Her Royal Highness visited the city three or four times subsequently; the last occasion was just before her death in 1786. She was a devotee of the gaming-table, was fond of fishing, horse exercise, and strong beer. As she grew in years she lost the slim figure of her youth, which did not trouble her in the least, and she was equally indifferent what people thought of her dress when engaged in angling. Her favourite haunt was a summer-house by the river-side in Harrison's Walks, where she often was seen attired in a riding-habit and a black velvet postillion cap tied under her chin. Her rubicund face and portly frame, with her novel headgear, must have made her general appearance more striking than elegant. When she took horse exercise she wore a hunting-cap and a laced scarlet coat, the latter perhaps in deference to the fashion of the time, which prescribed for ladies of all ranks a crimson riding-habit, faced with the colours distinguishing the regiment to which their fathers, husbands, relatives, or lovers belonged. Whatever her peculiarities, she was affable and generous, free from affectation, or any parade of her rank. She was partial to Mary Chandler, who kept a milliner's shop in the Abbey Churchyard, and gained some reputation with contemporaries for her poetical effusions. Her chief poem, "A Description of Bath," was dedicated to the Princess, who was also favoured with a special lyric, "in answer to Damon, who invited the nymphs of Bath to sing her praises." The first stanza will suffice as a specimen of its adulatory tone:

> Hark! Damon calls, I lead the way,
> Ye nymphs of Bath, come aid my lay;
>   Come, strike the trembling string;
> Amelia's name so sweetly flows,
> Her face such wond'rous goodness shows—
>   Who can refuse to sing?

The summer-house spoken of above must have been a building of some size and pretensions, as it is said to have had two fire-places in it. Possibly this was the grotto, so-called, which was afterwards found convenient for Sheridan and Miss Linley to carry on their clandestine correspondence, and which the verses of the former to Delia have immortalised.

A more illustrious visitor, and one to whom Nash paid assiduous court, was the Prince of Orange. The latter arrived in England in the autumn of 1733, with the view of marrying the Princess Royal; but having been taken ill, the ceremony was postponed. In the meantime the Prince settled in Bath, where he recovered his health. Nature had not been kind to him. He was short in stature and deformed. Lord Hervey states that he looked behind as if he had no head, and before as if he had no neck or legs. The Queen called him a monster; yet his face wore a pleasing aspect, and his manners were engaging. The Princess Royal doted on him; but the warmth of her affection was, it is said, coldly returned by her lord. To his presence in Bath we are indebted for two memorials. The Grove became the "Orange" Grove out of compliment to him, and the obelisk in the centre was erected at the expense of Beau Nash to commemorate the Prince's restoration to health "through the favour of God, and to the great joy of Great Britain, by drinking the Bath waters."

Four years later still greater fish came into the Beau's net. In the autumn of 1738 the Prince and Princess of Wales honoured him (or Bath) with their presence. They were enthusiastically received, and the great M.C. exerted himself to the utmost to make their stay enjoyable. The concourse of strangers was so great that the authorities found it necessary to regulate the price of provisions to prevent extortion. On the birthday of the King the Royal visitors were presented with an address; but as the relations between the monarch and his heir were then much strained the laudation of the former in the address must have made it difficult for the Prince to make a suitable reply. His portrait, as drawn by Lord Hervey, is repelling. He caballed against his father, insulted his mother, was unfaithful to his consort, untruthful, and very much of a blockhead. "He was true (says his lordship) to nobody and seen through by everybody." It is instructive to note what Smollett says of him in his continuation of Hume's "History of Great Britain." "This excellent Prince, who died in the forty-first year of his age, was possessed of every amiable quality which could engage the affection of the people; a tender and obliging husband, a fond parent, a kind master, liberal, generous, candid, and humane; a munificent patron of the arts, an unwearied friend to merit; well-disposed to assert the rights of mankind in general, and warmly attached to the interests of Great Britain." The only way to reconcile these contradictory delineations is to read Smollett's panegyric throughout in a negative sense. The truth is that the Prince was not as black as described by Lord Hervey, nor as bright as limned by the historian. His vices, no doubt, were redeemed by some virtues. Judging by his gifts while here the verdict would be favourable. Before leaving his Royal Highness not only cleared the prison of all debtors, but promised a thousand guineas to the fund for erecting the Mineral Water Hospital.

In July, 1750, the Prince and Princess again gladdened Nash and his lieges by paying another visit, being accompanied by the Lady Augusta, their eldest daughter. On alighting from their landau, at the lower end of the Market Place, they were carried in sedan chairs to the centre house on the South Parade. The inevitable address was presented by the Corporation, in the course of which their Royal Highnesses were congratulated on the birth of another prince. In the evening the Royal party drank tea at Ralph Allen's seat, Prior Park, and afterwards went to the theatre, to see the tragedy of "Tamerlane" performed by Mr. Linnell's company, at the command of Lady Augusta. Later in the week the illustrious visitors, attended by several noblemen and ladies, went down the river in wherries to Saltford, and dined in public under two tents in the meadow near the water side, where a great number of country people resorted, to whom the Prince gave about two hogsheads of beer. A band of music attended the whole time, and what added to the diversion (says the record) was the humour of the people, who danced several country dances, though as they were weighted with two hogsheads of beer the agility of some could have been no more conspicuous than their grace. Between eight and nine in the evening the Royal party returned in their wherries, the banks of the river being crowded with a cheering multitude. The next day their Royal Highnesses left Bath for Lord Bathurst's seat at Cirencester. In March of the following year the Prince died, and in the ensuing July a great cricket match was played in Saltford meadow, "in memory of his late Royal Highness having on that day twelvemonth dined in the said meadow, his Royal Highness being (it is added) a great admirer of that diversion, and it is intended annually to be followed there," which was not done. Perhaps the oddness of making an athletic game commemorative of a Prince's decease may have dawned on the projectors, and led to its abandonment, or their admiration of the defunct may have evaporated, there being no hogsheads of beer in view to keep it alive. The aquatic excursion of the Royal party shows how inviting were the lower reaches of the Avon in those days, with its waters uncontaminated by sewage and its banks fringed with willows instead of as now with houses and factories. It is a pity that no artist of the day made a sketch of the wherries and their gay occupants as they glided down the river.   Other and more permanent records of the Prince's visits the city possesses. There is the obelisk in the centre of Queen Square, which attests the homage of Nash to Royalty, by whom it was erected as an acknowledgment of the " benefit bestowed on the city by H.R.H. Frederick, Prince of Wales, and his Royal Consort in the year 1738."   The beautiful and costly loving cup which is passed round at civic banquets was also the gift of the Prince, as well as the full-length portraits of himself and the Princess which adorn the south side of the Guild-hall banqueting room. In the premature death of the Heir Apparent we probably see illustrated the truth of Shakespear's dictum --" The gods are just, and of our pleasant vices make instruments to scourge us."

Among other Royal visitors may be named the Princess Mary, fourth daughter of George II, and wife of the Landgrave of Hesse, who came in 1740, accompanied by her niece, the Princess Caroline. The latter returned in April, 1750, to use the Bath waters for a rheumatic attack. It is noted that "Richard Nash, Esq., our worthy friend and benefactor, went as far as Sandy Lane to meet and pay his compliments to her Royal Highness." A day or two after a fire broke out in the house occupied by her, she being away for a drive; it was subdued with the help of the soldiers quartered in the city, though the damage done was considerable. The night following the Princess slept at Mr. Nash's house. She appears to have lodged with a Mrs. Hutchings, close to the Baths, the water from which was used to supply the fire engines. Her Royal Highness gave ten guineas to be distributed among the soldiers, and ten guineas for the civilians who assisted at the fire. She left Bath on the 2nd June; but came again on the 24th November, and stayed till the 22nd December. The Prince of Hesse also spent some time here with his daughter, and received a visit from the Duke of Cumberland.   Ten years later the Duke of York, brother of George III., occupied the centre house on the North Parade; he gave a breakfast at Simpson's Rooms, and dined with Ralph Allen at Prior Park. The Duke of York, son of George III., and his Duchess were among the visitors in 1795, and the Prince of Wales the following year. H.R.H. during his stay was presented with the freedom of the city in a gold casket.

Frequently as the Princes were here, and more often than the foregoing record gives, the citizens were not satisfied. They anxiously expected the King and Queen, who were again and again announced to be coming. Due court was paid to their Majesties : portraits of both, life size, were placed in the Banqueting Room and Council Chamber, where they can still be seen, and other influences were brought to bear ; but all in vain. The King had two causes for disliking Bath. In the first place it had sent to Parliament as one of its representatives William Pitt (Earl of Chatham), for whom his Majesty had an invin-cible dislike, which he cherished with the obstinacy so characteristic of his narrow mind;

and, in the second place, Nash had the reputation of having given the city over to dissipation, which made it repugnant to his own undoubtedly high moral standard. He ever kept at a distance from it, though his aged Queen paid it a visit in the early part of the next century.

If these later illustrious visitors came under the cognizance of his successors, Nash had, it will be seen, a fair sprinkling of them to gild his reign. Its brightness, however, was sadly clouded at its close. In his early days the winnings he secured from the gaming table enabled him to live in fine style. The large mansion now known as "The Garrick's Head," if not built for him in 1720, was occupied by him for some years. His equipage was sumptuous; he usually travelled in a post chariot and six greys, with outriders, footmen, French horns, and every other appendage of expensive parade. With the curtailment of his illicit gains, he removed to the smaller house in St. John's Court. Though he was shorn of the emblazonry wealth enabled him to command, his power and authority remained undiminished as long as he was able personally to superintend the amusements. After he had passed his eightieth year his infirmities, mental and bodily, compelled him to relinquish part of his duties; in addition, his means were so precarious that he suffered considerable privation, which was partially relieved through the bounty of Ralph Allen of Prior Park, and at length more substantially by an allowance of ten guineas per month from the Corporation. Before this grant was voted his necessities compelled him to part with many of the valuable gifts he had received from distinguished visitors, including rings, watches, and jewelled snuff-boxes. He retained, however, an etui, mounted in gold, with a diamond to open it and another diamond at the top, given him by the Princess Dowager of Wales; a large gold snuff-box, enamelled, presented by the Prince of Wales; another from the Countess of Burlington, with Lady Euston's portrait on the lid; and a silver tureen, given by the Princess Amelia. He died in his 87th year, at his residence, and was buried in the Abbey at the expense of the city. His death gave birth to many elegiac pieces, and epitaphs, more or less suitable for his tomb, were composed by surviving friends, who gave unstinted praise to his services as M.C. and his kindness and generosity as a man. Yet great difficulty was found in raising the money for the monument in the Abbey. Even after the charge for the work had been reduced by the artist, Mr. John Ford, and the authorities of the Abbey had remitted the fees, there was a deficit of five guineas, which the promoters had to provide, so short-lived was the gratitude to him who, according to the circulars asking for subscriptions, was "a disinterested benefactor to this city, who first instituted its refinement, maintained its elegant police, and liberally supported every good work." The apathy is, perhaps, partly explained by the fact that several years elapsed before the proposed memorial was vigorously taken in hand. Immediately after his decease, Nash's effects were sold by auction; but the sum realised was not sufficient to pay his creditors in full. Later it was announced that all the family pictures in the possession of the late Mr. Nash would be sold at five guineas each, the vendor being "Mr. Yescombe, attorney-at-law." Like so many who take no thought of to-morrow, Nash in the heyday of his career squandered thousands, and died a dependant on bounty and insolvent.

SAMUEL DERRICK.

"We ne'er shall look upon his like again" could have been safely said of Nash by his contemporaries. He was an original of which there could be no reproduction. The author of the brochure of 1723, before mentioned, with keen insight, recognised that he was "sui generis."

Before the Beau's time none would have thought it possible (he says) that such absolute authority would have been allowed, " nor is it likely," he adds, " that after your decease it will ever be seen again." Apart from the combination of special qualities which gave Nash his distinguished isolation, the social conditions during his long reign of five-and-fifty years were undergoing a change, slow 'but sure, although his imperious will prevented any innovations to make the fact apparent. With his death came the period of transition, when the old order of things had to be modified. The " beau monde " asserted its supremacy, and began to regard the M.C. not as a master, but as a servant. It failed, at the same time, to realise the importance of having a man of tact, firmness, and experience to control the new spirit, and make it a success. Samuel Derrick, who (after a brief attempt to hold the post by Mons. Collette, the coadjutor of Nash in his later years) was chosen, had no recommendations for the office beyond those which certain literary achievements gave him. He was more familiar with the hacks of Grub Street than with " hautton," and if his talents and character won for him the esteem of Johnson and Boswell, his habits were never anything much better than Bohemian. With such a training he acquired a knowledge of mankind, but not the accomplishments for an " arbiter elegantiarum." Like many men who find themselves in an office for which they are ill qualified, he leaned upon others, and tried to please by giving ear to the advice of cliques, with the usual result of causing friction and discontent. Moreover, he was short in stature, which detracted from his dignity, and was labouring under financial embarrassments, the extent of which made it impossible for him to conceal his impecuniosity. He had to efface himself for some time, leaving Collette to play the part of warming-pan in the interval. At the outset his pen was busily employed in writing odes and sonnets to people of distinction who visited Bath, and these effusions, coupled with his good temper and amiability, won for him staunch friends, as his shortcomings nourished bitter enemies. Wrangling between the two sections was inevitable, and thus the seeds of discord were sown, which bore fruit after their kind a year or two later. The spirit rife is illustrated by an incident that took place at Wiltshire's rooms in January, 1763. At the Queen's Birthday Ball (records a private letter of the time) a little riot was worked up, " for the candles went out before eleven o'clock, the music went off in the middle of a dance, and left the company in the dark, who could neither get music, candles, nor even a little negus to drink, though they could prove the rooms cleared over five-and-forty guineas by subscriptions, upon which a gentleman said he remembered when such affronts as these were given it used to be the custom to break the lustres and glasses, upon which hint there was negus procured in plenty ; but the gentlemen threw it all over the room, broke eight bowls, and went off in a rage, swearing that there never should be another ball at these rooms. Wiltshire having made submission, they have passed it by and the balls go on as usual. Collette had carried himself off earlier upon some affront he had received, of which he has had plenty this winter, and since that night has resigned his office to one Derrick, a little Irishman, to whom, they say, the Rooms are to allow £50 a year. If that is the case it is no hard matter to prognosticate what authority he will gain and how far it will be attended to."

Notwithstanding the death of Nash and this undercurrent of dissatisfaction, Bath was more resorted to for health and pleasure than at any former period, a fact partly explained by the improvement of the highways under the various Turnpike Acts, and the better travelling secured. Thus in the autumn of 1765 as many as 148 persons of quality were numbered among the visitors. The list included three princes, four dukes, four duchesses, one marquis, two marchionesses, twenty-four earls, twenty-two countesses, fourteen viscounts, forty-three viscountesses, twelve barons, twelve baronesses, one ambassador, one archbishop, and five bishops. Among them were six Knights of the Garter, one Knight of the Bath, one Knight of the Thistle, and seventeen Privy Councillors. There was plenty of scope, therefore, for Derrick's adulatory pen; and likewise for the ringers to collect fees. As it was the custom for them to ring a joyous peal whenever a visitor of quality arrived, and as the line was not scrupulously drawn at the class thus designated, the tintinnabulatory din was well nigh continuous, and justified the protest made against the practice by the lovers of quiet, and particularly by invalids, who were distracted by the clanging noise. Anstey, with his customary light satiric touch, alludes to it in the "New Bath Guide":—

No city, dear mother, this city excels
In charming sweet sounds, both of fiddles and bells.
I thought, like a fool, that they only would ring
For a wedding, or judge, or birth of a king;
But I found 'twas for me the good-natured people
Rang so hard that I thought they would pull down
   the steeple;

So I took out my purse, as I hate to be shabby,
And paid all the men when they came from the
   Abbey.
Yet some think it strange they should make such a
   riot
In a place where sick folk would be glad to be quiet.

The justification pleaded for the custom was that it announced the arrival of new visitors and set others in residence inquiring who they were, a reason that failed to reconcile complainants to the hardship. When St. James's Church was being rebuilt the authorities, it is said, were offered £250 if they would not have bells. They refused, preferring the peal to the coin. Either before or after the ringers came the City Waits, who chanted or bellowed a roundelay in front of the lodgings of the visitors. For these "civilities" they expected a fee of a crown or half a guinea; the reward of the ringers was half-a-guinea, or a guinea, the sum varying with the rank of the people saluted.

Derrick's pen was also employed to enforce obedience to the rules of dress and decorum which Nash had framed. The repetition of these manifestos shows how difficult it was to get them obeyed. Ladies who intend to dance minuets are again and again reminded that they must appear in hoops, "without which they cannot but remember that no lady can be taken out for French dances." "Lappets" as well as hoops are imperatively demanded. "No lady can be permitted to dance a minuet without a lappet head and full-dress hoops, and such minuet dancers as choose to dance country dances will be attended in a retiring room by a woman servant to take their hoops off, as no hoops, be their size ever so small, are allowed in country dances." Gentlemen are likewise instructed in their sartorial duties. If they present themselves for a minuet they must appear "in full dress, or at least in a full-dress French frock suit," and if in leather breeches the latter will endure the same fate "as the country dancer's hoops, only no servant is provided to assist them." At his ball in January, 1766, "Mr. Derrick gives notice that no gentleman in weepers can be taken out for minuets on account of her Majesty's birthday."

The M.C. also tried to abolish the custom of giving subscribers' tickets to menials. "Owing to the very numerous company" he entreats subscribers not to give tickets "either to maid or manservant, as it is evident they are seated every night to advantage while people of very high distinction are obliged to stand." In those palmy days of Bath, society must have been much more tolerant, or stood less on its dignity, when the butler and lady's maid, the footman and cook were thought eligible to participate in the public pleasures of their masters and mistresses. The request for a little self-effacement on the part of the domestics and others who found their way to the Rooms was apparently treated with contempt. A few weeks later Derrick issued the following notice:—

"As the privileges of calling for chairs or benches in front is only due to persons of high quality, in order to prevent all future solicitations on that head, the Master of the Ceremonies finds himself under the necessity to observe to the public that no chair or bench can hereafter be ordered for anyone who does not rank as a peer or peeress of Great Britain and Ireland." With all the strivings of the M.C. it took some years of training before the "bench of ranks and degrees" was conceded and the promiscuous element kept at bay.

Derrick's health was already failing. After several attacks, he died at his house in the Orange Grove on the 28th March, 1769, being in his 46th year. In the following month it was announced that the subscription to the balls being much more than sufficient for the spring season, "it is proposed that part of the money shall go towards discharging the late Master of the Ceremonies' debts." A year later a public breakfast at Mr. Simpson's Rooms is advertised to take place on Easter Monday, "by particular desire of many ladies and gentlemen, after which there will be cotillons for the benefit of many persons who humanely supported the late Mr. Derrick in his illness and to the time of his death, to a considerable amount, and being in a low situation of life and not able to bear so great a loss, they humbly hope for the countenance and protection of the public." The accounts, it was stated, had been examined and their justice could not be doubted. They were probably due to tradespeople, who had supplied Derrick with necessaries during his illnesses; and considering how ruffled society was during his brief rule, the efforts thus made to discharge his liabilities were honourable and praiseworthy.

ANARCHY.

A contemporary writer describes the state of fashionable society in the preceding year as being in a state of anarchy. The epithet is scarcely fair, as a reasonable amount of order was maintained, though controversy was not silent. The anarchic was yet to come; the portents only had been seen. It was over the election of a ruler of the elegant and polite that the latent discord culminated, and that, too, in such scenes of disorder and violence as make the episode one of the most remarkable in the social annals of the century. Some weeks before the death of Derrick one Charles Jones, a well-known gamester, canvassed the nobility and gentry for their votes to place him in the post, and from a list of names published he obtained 185 promises, having given a pledge to "bid adieu to play." In the meantime, Mr. Brereton, a resident of Bath, officiated as the M.C., it being the intention of his friends to secure his appointment. Another disturbing element, however, appeared in the person of Mr. Plomer, who had been for sometime acting in a similar capacity at the Hotwells, and who aspired to fill the Bath post. The Bristolians suddenly became subscribers to get the right of voting for him, but the friends of Mr. Brereton would not concede the privilege, and promptly passed resolutions condemning the proceeding, and naming Brereton as Derrick's coadjutor and his successor in case of death. Mr. Charles Jones contented himself with protesting against both Brereton's and Plomer's tactics, while a fourth candidate, Captain William Wade, nephew of the late Field Marshal Wade, member for the city, withdrew from the field with a promise that he would come forward at the next vacancy. He was rewarded for his becoming conduct, as the sequel will show. While Derrick was yet living the friends of Brereton proposed to buy off Plomer by promising £100 a year out of the subscriptions. The next day Derrick died, and in the course of the afternoon Brereton was formally elected Master of the Ceremonies. The Plomerites affirmed that the election was illegal. They held a meeting at the Town Hall under the presidency of Lord Clare, when a resolution was passed declaring that three days' notice (as before intimated) was the shortest that would be expected by the company to render any meeting for the election of a Master of the Ceremonies valid, and appointing a day for such election; declaring further that their rights as subscribers to the balls had been invaded by the ladies and gentlemen who at Gyde's (late Wiltshire's) Rooms had chosen Mr. Brereton. The battle waxed furiously; charges and counter charges were made in public and in private against the candidates by their opponents; pasquinades also added fuel to the flame. The friends of the Bristol candidate fixed Friday, the 31st March, as the day of election at Simpson's Rooms. He, in his address, expressed himself "gratefully sensible of the repeated honours conferred on him by subscribers to the Great Assembly"; but he had not yet received his chief distinction. This was vouchsafed to him on the eve of the election. Both factions had gathered at Simpson's, and after a wrangle and some hustling Mr. Plomer was led out of the room "by the nose," despite the desperate efforts of his friends to release his nasal organ from the Breretonian grip. The attack was brutally rude, but its singular nature makes it, with the lapse of time, irresistibly comic. Happily, he was a man of peace, and yielded to the wish of his supporters that he should not resent the insult in the way then deemed becoming gentlemen. He was in due course elected, and held his assemblies at Gyde's Rooms, while his rival ruled at Simpson's. A squib issued at the time thus alludes to the unpleasant incident just narrated:—

The time elaps'd, the books they close-a,
Declaring Plomer duly chosen-a,
Which was a balsam to his n—e-a.
  Doodle doodle doo.

To heal the difference, what hope-a!
While one side talks of nose and soap-a,
And t'other threatens with a rope-a.
  Doodle, etc.

Then anger, wrath, and party zeal-a,
Loud bawling for the public weal-a,
Made laws and government to reel-a.
  Doodle, etc.

In their zeal for their favourite candidate each party represented the other as gamblers and fortune-hunters on one side, as tallow-chandlers, cheesemongers, rope-makers, etc., on the other, while the decencies of private conversation were violated in tales of malice and calumny whispered abroad. With society in this inflammable state it needed only a spark to cause an explosion. It was supplied at the Master of the Ceremonies' Ball at Simpson's on the 3rd April. Before the minuets began a gentleman, in the interests of Mr. Plomer, wished to read a written protest "against any person whatsoever who shall presume, by violence or otherwise, to act as Master of the Ceremonies," but hisses, groans, and indecent remarks from the other party made the attempt abortive. Then a Mrs. Hillman, described as a "parson's fierce termagant wife," contended for precedency with a "noble peer's daughter," and was knocked literally head over heels. A scrimmage ensued, with the characteristic violence of an Irish faction fight. The gentlemen fought and swore, the ladies, screaming, tore each other's garments and headgear; the floor was soon strewn with fragments of caps, lappets, and millinery, coat-tails and ruffles; whilst the injured, with handkerchiefs to their damaged faces, strove to extricate themselves from the mêlée. The sons of Mars "a bloodier battle wage;" and tall heroes lay upon the floor dyed as it was with purple streams and encumbered with scraps of finery. The non-combatants hurried to the exits, or mounted the chairs near the walls to be out of danger or to watch the foes mauling and bruising each other. So the conflict raged; its fierceness paralysed intervention; and the help of the civil authority was at length sought to restore order, which was not secured until Jupiter descended in the person of the Mayor, armed, not with the dread thunderbolts, but with the Riot Act. Even his Worship's efforts, aided by the Town Clerk, though strenuous, were for some time futile, and the Act had to be read three times before the tumult ended, though savage threats and oaths, like the final rumblings of a thunderstorm, prolonged the confused din. Probably the fray was intensified from the national passions engendered, Mr. Brereton being an Irishman and Mr. Plomer an Englishman.

CAPTAIN WADE.

Such a scandalous exhibition of folly and fury under circumstances so unprecedented was soon noised all over the country. Wags made merry over it, and revelled in the details, which lost nothing by carriage; but the thoughtful reluctantly believed and lamented the degradation, self-inflicted, of a fashionable assembly. Meanwhile, the lovers of peace and order lost no time in trying to bring about an amicable arrangement between the contending factions. After a good deal of negotiation a compromise was effected, the chief articles of which were that both Mr. Brereton and Mr. Plomer should retire; that Captain Wade should be the Master of the Ceremonies; that the proprietors of the different rooms should obey him in that character; that the sum of five hundred pounds, part of the surplus money subscribed to the balls during the season, be given to Mr. Brereton and the remainder to Mr. Plomer, with a benefit ball for each; and that Mrs. Brereton should have a ball annually for her benefit in the month of November. Thus ended a memorable struggle, which was something more than a nine days' wonder, for it lived vividly in the memories of those who actively engaged in or were spectators of it.

A fitting seal was set to the treaty of peace by the presentation to Captain Wade of a gold medallion, to be worn by him as an honorary ensign of office. It was of fine gold, enamelled blue, and elegantly enriched with brilliants. On one side was a raised figure of Venus, with a golden apple in one hand, and a rudder in the other: the motto, "Venus decens." The reverse was a wreath of laurel, and the motto, "Arbiter Elegantiæ, communi consensu." This interesting souvenir was lost, possibly in the great fire at the Lower Rooms in 1820.

The social harmony secured by the means described did not, however, last long. In the autumn of 1773, the new Assembly Rooms having been opened, a feud arose between the proprietors of these and the lessee of the old rooms as to the time of holding the amusements at the respective places. Society took sides in the dispute, and a rancorous controversy, extending over weeks, was the result. When Captain Wade was elected it was understood that the control of the amusements would be left entirely in his hands. On the completion of the new rooms he had an interview with the proprietors, and accepted from them the post of M.C., subject to a condition that they should have a voice in the management. The compact did not work satisfactorily, as the programme of the upper rooms was found to clash with that of the lower. Wade favoured the latter, or was thought to do so, and the proprietors of the former censured him for being remiss in the discharge of his duty. Lord Camden, Lord Cadogan, Lord Southwell, and other notables espoused Wade's cause; Mr. Claude Crespigny, with several of the nobility, stood by the new rooms. Meetings were held by both parties; overtures and counter-overtures were made to bring about a settlement, but the labour was in vain; the dispute went on. The proprietors fixed balls and concerts when they liked; the M.C. did the same at the old rooms. The bitter spirit evoked may be inferred from the language employed and the reprisals resorted to. "With indignation suitable to the occasion" was the prelude to the rejection by Lord Camden and his friends of the terms proposed by the opposition. Lord Camden, on the other hand, was accused of making use of his honourable leisure to become a "Bath brawler," and was credited with the intention of getting an Act of Parliament passed to compel the proprietors of the new rooms to enter into partnership with the keeper of the old ones, his action being delayed until his son had been returned member for Bath—"to gain which point a truly respectable character is descending very low indeed." Such were the kind of calumnies freely circulated, and notwithstanding the absurdity of some of them, readily believed. Being accepted as fact, they superheated a controversy already fierce enough without these vitriolic additions. From the other side we learn that "at the last cotillon at the upper rooms, except a few antiquated beaux for partners, the ladies (in duty to the old folks) danced with each other." Machines to warm the rooms are promised "lest any should catch cold from the thinness of the company." Wade, on his part, peremptorily ordered the band "to attend the ball on Monday at Gyde's rooms on pain of being discharged," although the unfortunate musicians were under engagement to play at the upper rooms. Subscribers to the latter were likewise declared incapable of subscribing to the former, in the hope that by these coercive means "peace and harmony would be restored." All this factious rage arose over the issue whether there should be a consolidation of the amusements, as Lord Camden desired, or independent management, for which the proprietors contended. To dispassionate minds, it seems obvious that the control of the amusements at both sets of rooms by an M.C. holding the scales evenly between opposing interests was more likely to promote concord and prosperity than a system of rivalry, with each trying to best the other; but the proprietors would not see the question in this light, thinking that the great cost incurred in erecting the rooms entitled them to special privileges. Richard Nash, "from our card-room in Shades below," was supposed to send them the following friendly expostulation :—

Ye Rulers of the Upper Regions,
You must, my friends, submit to legions;
In ev'ry state 'tis the majority
That claims the sovereign authority.
To numbers shall the few give law.
Can Mahomet the mountain draw?
In vain your stately rooms you boast,
D—mn ye, you shall not rule the roast.
Your balls, your concerts, your cotillons,
Are here below, outdone by millions.
Then do not, by your ill-timed wrath,

Destroy the harmony of Bath,
Where Invalids, both small and great,
Expect an indolent retreat;
Not to be plagu'd with your damn'd jangling,
Like lawyers at a sessions wrangling.
Don't you perceive you've thinn'd the season?
Unless then you'll submit to reason,
And be content with moderate gain,
Your vapours will descend in rain;
All future company will resent it,
And you, my friends, too late repent it.

A conviction of the folly of continuing the strife at length leavened the opposing bodies, and paved the way for a settlement. Captain Wade, who had been censured by the proprietors for exercising despotic authority, made an explanation or apology, which the former accepted as satisfactory, more particularly as they were to be duly apprised of his arrangements prior to their being carried into effect. He was, therefore, left "sole director of the diversions"; but he soon found that his authority was not absolute. Having made alterations which were not acceptable he had to waive his right, and cancel them. At the spring season of 1775 he took "the liberty to recommend the following amusements" for each week :

Monday.—The Dress Ball at the New Rooms.
Tuesday.—Public Tea and Cards at the New Rooms.
Wednesday.—The Cotillon Ball at the Old Rooms.
Thursday.—The Cotillon Ball at the New Rooms, and Tea and Cards at the Old Rooms.

Friday.—Dress Ball at the Old Rooms.
Saturday.—Public Tea and Cards at the Old Rooms.
Sunday.—The Rooms to be open alternately for Tea and Walking. The Old Rooms to be opened on Sunday next.

It will be seen from the programme that the claims of the old and new rooms had fair recognition, and all cause of antagonism being removed, peace and amity reigned instead of discord and jealousy. The only time when a renewal of the strife was threatened was at the resignation of Wade. Several candidates again presented themselves, and a conflict between the old and new rooms, or their patrons, was imminent. It was averted, however, by an agreement to have a Master of the Ceremonies for each place, it being considered desirable "because the city was now so large and the resort of company to it was so great."

WADE'S EXIT AND HIS SUCCESSORS.

Wade, both in storm and calm, aimed at conducting the amusements on the lines laid down by Nash. The great Beau's system of administration was regarded as a standard of perfection, which it was the duty of all to maintain. Fashions might change, tastes alter; but the code of Nash was treated as fixed and unalterable. He was one of the "dead yet sceptred sov'rans who rule us from their urns." Still it was by no means an easy task to reconcile the younger generation, who knew Nash only by name, to the strict discipline as to dress and deportment that he was able to enforce. There were rebels in the matter of toilettes, whose self-will and defiance sorely tried the M.C., as it did his predecessor. No one ventured knowingly to infringe the regulations of Nash; a look or a hint from him would abash the offender. He was "monarch of all he surveyed," and people knew that ostracism from the assemblies was the doom awaiting the refractory. The deference thus exacted and readily given no longer smoothed or lightened the cares of office.

There was an undercurrent running in the direction of relaxation. Smollett voiced the demand when he complained of the "fatigue and slavery of maintaining a ceremonial more stiff, formal, and oppressive than the etiquette of a German elector." A jaundiced critic (Thicknesse) declared that "the dancing at the rooms had all the solemnity of a funeral." Even with this dissatisfaction latent it was very difficult to obtain any modification in the routine prescribed or the habiliments to be worn. The fact is illustrated by the struggle to get the "night-gown" (evening dress as now called) recognised as suitable wear. In 1772 Wade proscribes it "in toto"; a year or two later he permits it, as "the French night-gown," to be worn by ladies who did not intend to dance; under his successors, in 1777, the "Italian night-gown" is received into full favour. "What's in a name?" Much when it covers a concession. Evidently the pertinacity of the ladies won the day against the sticklers for the Nash ideal. An idea of the garment which successfully ran the gauntlet of this opposition may be gathered from a letter written by the Duchess of Devonshire to her mother in February, 1785: "My sister and I," she says, "were very smart for Carlton House. Our gowns were night-gowns of my own invention: the body and sleeves black velvet, bound with pink and fastened with silver buttons; the petticoat of light pink and the skirt, apron, and handkerchief crape, bound with light pink; large chip hats with feathers and pinks."

Wade did his best, as stated, to preserve the old régime, but his manifestos assume a suppliant instead of a dictatorial tone. The only occasion when he adopts a more severe style is in the autumn of 1772. The edict then issued begins, "Whereas it was observed at the ball on Monday night last that several persons did appear improperly dressed," etc., and concludes that "for the future any person wearing a night-gown or apron will not be admitted of a ball night." As a rule, he "humbly requests" compliance with the regulations. "It is humbly requested that those ladies who do not dance minuets will not take up the front seats at the balls, except ladies of precedence." Another evil which Derrick vainly strove to abolish still flourished as the following stern notice shows: "Servants, hairdressers, and other improper persons who every night occupy some of the best seats, and even presume to mix with the company, are warned to keep away, and so spare themselves the mortification of being desired to withdraw, a circumstance which will inevitably happen if they continue to intrude themselves where decency, propriety, and decorum forbid their entrance." The reason for the efforts to exclude all novelties is to be found in the decree declaring that "it would be utterly impossible to preserve any propriety in dress if the rules of this assembly were to vary with every trifling alteration of the fashion in London and Paris." To prevent mistakes, ladies who choose to dance minuets at the dress balls "are apprised that a suit of cloathes, or a full trimmed negligès, with lappets, are the only dresses proper for the occasion; that all other fancy dresses, such as polinèse, French night-gowns, etc., however elegant, are highly improper to be worn with lappets." Ladies who may have just returned from the courts of Versailles or St. James, bringing with them the newest fashion, must have been deeply chagrined, when instead of being "the cynosure of all eyes" as expected, they had to lay aside the special attraction and become inconspicuous units in the crowd. The Queen of the West sat apart; she prescribed her own rules as to the attire suitable for the votaries of pleasure, and her judgment in such matters could not be swayed by the caprices of other centres of fashion either at home or abroad. An example this of Chinese exclusiveness on a small scale.

These pretensions received a rude shock when the fashion of wearing high plumes of feathers on their heads was all the rage with the ladies, the absurdity of which the caricaturists of the age did not fail to exhibit. It was introduced from France, and the finery was as common in Bath as elsewhere, giving rise to the following lines:—

Capricious, airy, feather'd race?
(For sex, alas! is fled),
Say what has mortaliz'd each grace
And cockatoo'd each head.

Can nodding plumes, the warrior's meed,
Give softness to the eye?
Or think ye Cupid is decreed
To take his stand so high!

To Gallia then return this toy,
Gallia who sent it hither,
Lest Fame might tell this truth with joy
"Each head much lighter than its feather."

The affection for hoops disposed the fair to cling to them in season and out of season, and resort to various expedients for the purpose of eluding the vigilance of the M.C. It was again and again ordained that no lady should dance in country dances in a hoop of any kind. The practice came up in another form, as it is officially announced that "hoops of the smallest size, commonly called pocket hoops, are by no means proper to be worn with lappets. It is therefore expected that every lady who chooses to dance minuets will wear a hoop suitable to the fashion and proper for the occasion." The regulation dress in 1777 was "a suit of cloathes and full trimmed sacque, or full trimmed Italian night-gown and petticoat, with lappets and dress hoops."

The vagaries of the gentlemen in dress were kept, as of yore, in check with the same rigour. "Costly thy habit as thy purse can buy, but not expressed in fancy," was the general law to secure uniformity. They were forbidden to wear "fancy dresses with slash sleeve or jacket, or turn-down cape," etc., the only dress proper for minuets being "a full trimmed suit of cloathes, or French frock; hair or wig dressed with a bag." Any other costume was "insufficient to attend the ladies who are obliged by the rules of the assembly to appear in full dress." Neither were they allowed to wear spurs in the Pump

Room in the morning any more than boots in the public rooms in the evening. A preamble to the regulations pleads for their strict observance on the ground " of its being absolutely necessary that the utmost decorum in manner and propriety in dress should be observed in so polite an assembly as that of Bath." About this time a waggish trifle caused some amusement. Ladies in their leave-taking letters were wont to append the letters " A.M.C." (" adieu, mon cher "). Sundry missives were sent and received with " D.I.O." substituted, the meaning whereof was a riddle no one could readily solve. At length it oozed out that the cryptogram stood for " D—me, I'm Off ! "—a formula adopted by the fast spirits of the day. Of more serious import was the prophecy in which Lord Mansfield at this juncture indulged. The Duke of Northumberland remarked to him one day, " There is one comfort I cannot have at Bath. I like to read the newspapers at breakfast ; but here the post does not come in till one o'clock. That is a drawback to my pleasure." " So," said Lord Mansfield, " your Grace likes the comfort of reading the newspapers. Mark my words, a little sooner or later these newspapers will most assuredly write the Dukes of Northumberland out of their titles and possessions, and the country out of its King. Mark my words, for this will happen." The doom thus confidently foretold by the Tory Chief Justice has not yet come to pass, and the nation can still take comfort in the fact that there is no prospect of its fulfilment.

The official career of Wade, whose portrait by Gainsborough still adorns the Assembly Rooms, came to an abrupt termination in July, 1777. He was involved in an affair of gallantry, and as the result of the action at law he had to pay the injured husband a solatium. His resignation was tendered and accepted ; in the struggle impending for the election of his successor he unwisely offered himself as a candidate, but only to learn that he was not wanted in Bath, acceptable as he was at Brighton. As previously mentioned, an amicable feeling was secured by the resolve to have two Masters of Ceremonies instead of one. The gentlemen chosen were Mr. Dawson and Major Brereton, the former officiating at the Upper Rooms and the latter at the Lower Rooms, now centred at Gyde's. A second medallion was then deemed necessary, and such was presented to Mr. Dawson. It is of fine gold, enriched with brilliants ; one side displays the figure of Minerva with the motto " Decus et Tutamen " over it, and under " Dulce est despere in Loco," a rich border of enamel encircling the whole. On the reverse is inscribed " Arbiter Elegantiarum " in a wreath of laurel and palm, beautifully enamelled on a blue ground. At the end of three years Major Brereton resigned, and was succeeded by Mr. Tyson, from Tunbridge Wells. In 1785 the latter was translated to the Upper Rooms on the retirement of Mr. Dawson ; and Mr. J. King, who had highly distinguished himself in the British Army during the American war, was unanimously chosen for the Lower Rooms, the two ·continuing to act for the remainder of the century. During the reigns of these several officials the old arrangements for the balls, concerts and cards, and the attendances at the baths and Pump Room, were generally maintained ; but the question of dress was, as at earlier periods, a recurring cause of friction between the M.C. and the votaries of fashion, who declined to be bound in matters of detail to the sumptuary edicts of the directors of amusements. In these struggles the ladies usually came off the victors, and thus modifications were introduced. In 1784 King Dawson issues a decree on hats : " Ladies are humbly requested to observe that the new-fashioned hats, commonly called ' balloon hats ' (as they are contrary to the established rules of Bath) will not be admitted into the rooms on any evening." Despite this positive prohibition there was a growing disposition to be more tolerant of innovations, as may be gathered from a manifesto issued by his brother monarch, Mr. King, in the following year. "It seems," he says, "perfectly reasonable that the regulations of dress should be conformable to the prevalence of fashion, and accommodated to the pleasure of the ladies as far as the popular customs of Bath can warrant. Upon this principle the Master of the Ceremonies, anxious to show every mark of respect and attention to the ladies, thinks it advisable to indulge them in their desire of coming to the rooms on Wednesday and Sunday evenings in hats that are expected to be in some degree ornamental, otherwise it will induce the necessity of prohibiting them: But it must be observed that hats of no sort will be allowed either at the concerts or cotillon or dress balls; and should any lady, through inattention, or any other motive, infringe this regulation, she must not take it amiss if she should be obliged to take off her hat or quit the assembly."

Arduous as were the duties of the Masters of the Ceremonies, it cannot be said that their remuneration was excessive. After the death of Beau Nash each had a benefit ball in the winter and the spring season; subscription books were also laid down at the Rooms, "that such of the company who were not present at the balls may have an opportunity of showing those gentlemen marks of their respect."

If dressing was expensive the charge for the amusements was extremely moderate. A payment of one guinea entitled the subscribers to admission to all the dress balls (twenty-six in number); half-guinea to all fancy balls (thirty in number). An extra sixpence was paid at the doors, for which tea was provided. There were nine subscription concerts, and three choral nights in the winter at the New Rooms, under the direction of Signor Rauzzini. A subscriber of two guineas had an admission ticket for each concert, and two tickets transferable to ladies only. The theatre was open on Tuesday, Thursday, and Saturday. The Rooms were also open every day (Sundays excepted) for cards, and every other Sunday evening for a promenade and a sixpenny tea. It may seem remarkable to moderns how human nature sustained the strain of such a constant round of amusements. An explanation is supplied by the fact that dancing was limited to four or five hours, and that it terminated generally at eleven. Grace and stateliness, too, were its characteristics; it was not hurried, but leisurely, consequently it was less exhausting, while it admitted of a freer participation by seniors in the pastime than the popular dances of later times allow. The band, it was stipulated, should consist of twelve performers, and include "an harp, tabor, and pipe." The fee to each musician was not to exceed half-a-guinea on each night.

Such, then, is a review of the gaieties of Bath through the century. The order and discipline generally maintained exercised, as already hinted, a salutary influence on the habits and manners of an age more noted for its licentiousness and coarseness than for its propriety and self-restraint. The long reign of the régime is the best proof of its beneficial tendency, as well as the countenance it received from ministers of state, judges, generals, bishops, poets, and philosophers, who made Bath a rendezvous. The benefits it conferred should indeed secure for it a kindly remembrance, however peculiar, in the light of the present, may seem some of its characteristics. Neither was it a mere vain boast a century ago and more which asserted that the local entertainments "are so wisely regulated that, although there is never a cessation of them, there is never a lassitude from bad hours or from an excess of dissipation." Moreover, of the visitors who crowded the city it could not be said that they took their pleasure sadly. There is a sprightliness in the "vers de société," "jeux d'esprit," and other ephemeral productions of the pen, which indicates the prevalence of a buoyant temperament and a robust participation in all the good things going. The pessimistic question " is life worth living?" would doubtless, if asked in this realm of Euphrosyne, have been regarded as an inquiry emanating from a victim of mental aberration. It is amidst the fever and the fret of the Plutocratic civilisation of the present that the question is not only asked, but seriously discussed, and the light-heartedness of our ancestors envied.

"The Entrance Door
Beau Nash's House

CHAPTER III.

THE BATHS AND BATHING CUSTOMS.

HE allusions previously made to the Bathing establishment have given some idea of its condition and management by the authorities, but without more details the knowledge of the reader would be very imperfect. It is these that the following pages will supply. The Baths were five in number, chief in importance being the King's Bath, with its adjunct, the Queen's Bath, situated in the rear of the Abbey. The spring, as now, rose in the centre of the former, and was covered with a leaden cistern to restrain its rapid flow and to disperse the water more equally throughout the area. A wooden building stood in the middle surmounted by an octangular tower. It was 64ft. in length and 41ft. in breadth, and had recesses on each side for the bathers to sit in; those on the east and west sides were called "The Kitchen," from the excessive heat of the water in that part of the bath. At the four corners were slips opening to the bath for the bathers to disrobe and dress in, with steps leading to the water. On the north side was the Pump Room, and on the south and west were the lodging houses, the east end being more open, as there was a passage between the margin of the Baths and the Priory, which completed the tenemental frame. Terraces, free to the public, surrounded the Baths, giving a view of the bathers. A jug was used to dip up the water or catch it from a pump, and thence poured into a long-stemmed glass for the invalid to quaff, if ordered to take it internally. There was no other protection from the elements or from the gaze of spectators who lounged around than the niches or seats. Thus exposed, it was no wonder that visitors complained of the risks and inconvenience they endured, which was the case with Sir William Thompson, one of the Barons of the Exchequer, who told "the Mayor and his brethren that he, by a line of wind from a low corner of the walls surrounding the King's Bath, had catched such a cold as had like to have caused his death." The Corporation listened to complaints but did nothing. Inducements could not tempt them. The Earl of Marlborough once offered to roof over the Cross Bath if the authorities would similarly protect the King's Bath. The offer was declined, as the cost of the Corporation's share of the work would have been much greater than that of the Earl's.

The Cross Bath appears to have received a little more attention, it being made use of by people of fashion to bathe in for pleasure. It was surrounded by a high wall and had a small gallery at each end for the band, while from the centre rose a cross of fine marble, which included, among other embellishments, the descent of the Holy Ghost attended by angels, and shells borne on the heads of cherubims. The cross with a crown of thorns surmounted the structure, which bore likewise several inscriptions, their purport being to show that it was erected by the Earl of Melford, Secretary of State to James II., as a memorial of the Queen bathing in it in 1687. As it was a Stuart who was honoured with this monument, it was viewed with aversion, especially during the rebellion of 1715, when it was much defaced; fanatics saw emblems of idolatry in the Cross and Holy Ghost, and tried to efface them. In course of time this extraordinary structure became unsafe, and it was cleared away some years before the bath was rebuilt by Baldwin.

The Hot Bath (so called from its superior heat), was fairly protected from the wind, and had a cross or tower in the centre, also three slips or cells, and thirteen arches against the walls for shelter. The Act of Parliament under which the Mineral Water Hospital was established gave the inmates of that institution the right to bathe in it, but the privilege became obsolete when the mineral water was conveyed to the Hospital and proper baths were there provided. The cross, niches, and cells disappeared when the younger Wood carried out the extensive alterations made under his auspices in the last quarter of the century.

The fifth, or Leper's Bath, adjoined the Hot Bath, from the overflowings of which it was supplied. Its area was no more than ten feet by eight feet, until enlarged by Wood, and had only one small slip. The inmates of the Leper's Hospital—a house so called near at hand—had the exclusive use of the bath, the efficacy of which was attested by the following inscription on the side of the cistern:—"William Berry, of Gatharp, near Melton Mowbray, county of Leicester, cured of a dry leprosy by the help of God and this Bath, 1737."

As in these smaller baths, so in the much more important King's and Queen's Bath, all the objectionable features above recorded were swept away by the grand scheme of reconstruction effected by Baldwin, while other improvements introduced made them as convenient and comfortable as they were before mean and dangerous.

The Corporation was no doubt stimulated to renovate the baths by the example set by the Duke of Kingston. When the Abbey House (his Grace's property) was pulled down in 1755, several stone coffins and Saxon coins were found beneath. Lower still, extensive remains of Roman Baths were brought to light, including hypocausts, sudatories, and frigidaria, also what was thought to be another hot spring. The existence of the latter induced the Duke to utilise it, and he accordingly built three private baths with domed vaulting and steps into them from distinct dressing rooms, but to make room for these and the foundation of other property nearly all these vestiges of antiquity were ruthlessly swept away. It was surmised at the time that they indicated merely one wing of a magnificent building, a surmise which later discoveries have confirmed, although even now the whole area has not been explored. The good example the Duke set the Corporation in providing private baths was, however, dearly purchased, when it involved the sacrifice of far more interesting details of the Roman balneal system, judging by the descriptions left of them, than any that have since been found. The so-called spring, which suggested the Duke's enterprise, proved ultimately to be simply a leakage from the King's Bath; the water percolated through the intervening ooze, and was intercepted at the spot where his Grace turned it to profitable account. In fact, there are only three springs; one supplies the Grand Pump Room, the King's public and private baths, the baths in the Royal Mineral Water Hospital and the Pump Room Hotel suite of baths; the second supplies the Cross

Bath, and a third the Hot Bath, the Royal private baths, and a bath in the Royal United Hospital. The three were probably known to the Romans, the only doubt being with regard to the Hot Bath spring, which, unlike the other two, was not protected at its source by a massive stone chamber of Roman construction. The leakages from it were in consequence excessive until the younger Wood confined it within a cylinder of stone, as previously described.

The mode in which bygone generations used the baths reveals some curious customs. Not the least singular to modern notions was the bathing of the sexes together. The practice was often condemned; but it was difficult to abolish. An attempt was made in 1753 to reform it, but the opposition the proposal encountered compelled its withdrawal. In Anstey's time, as we learn from the "New Bath Guide," men and women still bathed together, and it was not till quite late in the century that a separation of the two was effected. The dresses worn by the bathers gave scope for coxcombry and coquetry, and probably strengthened the resistance to change. In the major part of the century there were regulation dresses. The ladies wore garments of fine yellow canvas, large and stiff in quality, "with great sleeves," says Celia Fiennes, "like a parson's gown; the water fills them up so that your shape is not seen. The gentlemen have drawers and a waistcoat of the same material; this is of the best linen, for the bath water will change any other yellow. You generally sit up to the neck in the water, unless you required to be pumped on, as some persons are, for lameness, or on their heads for palsies. I saw one pumpt (she adds); a broad-brimmed hat, with the crown cut out, was put on so that the brims cast off the water from the face. The water is scalding hot from the pump. The arms or legs are more easily pumped. There are galleries round the bath where the company, who do not bathe, may walk or look over into the bath on their acquaintances and company. Guides are attached to each bath—women to wait on the ladies and men to wait on the gentlemen, but they keep their due distance. There is also a sergeant, who all the bathing time walks in the galleries, and takes notice that order is preserved, and punishes the rude. Most people of fashion send to him when they begin to bathe; he takes particular care of them, and compliments them every morning, which deserves its reward at the end of the season."

"When you go out of the bath" (continues the same lady) "you go within a door that leads to steps still in the water, and as you ascend the door is shut. You ascend several more steps and let the canvas fall off by degrees into the water, which the guide picks up. In the meantime your maids fling a garment of flannel (made like a night gown with great sleeves) over your head; so you are wrapped in flannel, with your night gown on the top, and having put on slippers, you are set in a chair which is brought into the room (called slips). The chair has a low seat, with frames round and over your head, covered inside and out with red baize, with a curtain in front of the same material, which makes it close and warm. Then a couple of men with staves takes and carries you to your lodgings and set you at your bedside. Then you go to bed and lie and sweat some time, as you please." One drawback to this mode of conveyance the gossiping Celia does not mention, and that was the occasional manifestation of temper by the chairmen, at whose mercy the fare was. If they were not fee'd as liberally as they expected, they would put the chair down, lift up the top or draw aside the curtain and expose the victim to chilling blasts of air, remonstrance being met by some paltry excuse. Some of them, too, were not remarkable for sobriety, their unsteady gait causing no little alarm to the lady or gentleman of whom they had the charge. It is recorded of one elderly matron that after being tossed and jolted about sufficiently to cause acute nervous apprehension for her safety, a halt at length took place, when she was severely rebuked for her restlessness, and told that if she did not sit still she would not be able to get to her lodgings!

According to Wood, the linen worn in the bath was originally white, but acquired its saffron hue from the waters, and was not purposely of yellow colour as Celia Fiennes imagined. Ladies who bathed for pleasure in the Cross Bath, were enabled to pass the time very agreeably. "In the morning (wrote Defoe in his "Tour through Great Britain") the young lady is brought in a close chair, dressed in her bathing clothes to the bath. There the musick plays her into the water, and the women who tend her present her with a little floating dish, like a bason, into which she puts an handkerchief and nosegay, and of late a snuff box is added. She then traverses the bath, if a novice, with a guide, if otherwise, by herself; and having amused herself near an hour, calls for her chair and returns to her lodgings." It is said that a young lady so engaged, seldom failed of becoming an object of admiration to some young gentleman in the gallery at the side of the bath, or of receiving those compliments which a fine glow of countenance, arising from the heat of the waters, must necessarily draw from admirers.

Anstey found in the Baths and their use a subject well suited to the ironical vein which the "New Bath Guide" reveals, but without adding anything new to the descriptions previously given. In his rattling way, he says:—

"You cannot conceive what a number of ladies / Were wash'd in the water, the same as our maid is. / Oh! 'twas pretty to see them put on their flannels, / And then take to the water like so many spaniels. / 'Twas a glorious sight to behold the fair sex / All wading with gentlemen up to their necks."

The ladies, full toiletted and "bien coiffees," were, we know, escorted by their cavaliers, with powdered hair and bagwigs, and indulged in all the luxury of the bath at a temperature of 105 degrees.

The waters were drunk from six in the morning during the summer season, and seven in the winter, until ten, while the chief time for bathing was between six and ten. The quantity taken internally was large, being from three pints to two gallons per day. Dr. Cheyne prescribed five pints. The fires in the slips were lighted at five in the morning during summer and six in the winter. The band performed in the Pump Room from eight to ten for the entertainment of the company, who made a very gay appearance. Mrs. Schimmel Pinninck, in her autobiography, describes the scene revealed to her young eyes in 1788—the centenary of the glorious Revolution of 1688. The occasion was used to show the staunch adherence of the nation to Whig principles and its dislike of the personal rule George III. was trying to introduce. Not a lady was to be seen in the Pump Room without streaming orange-coloured ribbons, nor a gentleman without rosettes of the same in his buttonhole. Balloons were at that time just come into vogue, "and everybody wore huge balloon bonnets with magnificent ostrich feathers," other noticeable features in the attire being "the ample muffs and long tippets with fur linings of the silken Angora goat's hair." It was a brilliant picture, teeming with colour and vivacity, while above the conversational din heart-stirring music resounded from the orchestra.

Turning to the medicinal virtue of the waters, one is surprised at the remarkable cures which they are credited with effecting. Proof thereof is not only to be found in the works of the faculty, but in the substantial marks of gratitude left behind by the patients restored to health. Whether the diseases from which they were suffering were in all cases correctly diagnosed is a point on which subsequent experience has thrown considerable doubt. In any case, many of "the ills that flesh is heir to" were removed, and sufferers went away rejoicing, first testimonialising the springs to which they were so deeply indebted. Sir Francis Stonor received such an extraordinary relief from "gout and aches" that he gave the Corporation a sum of money to pave the passages on the north, east, and west sides of the King's Bath, besides surrounding it with a carved stone balustrade, a portion of which still remains. Robert, Lord Brooke, having been cured of a diabetes, erected the minstrels' gallery in the Cross Bath, and placed a handsome chimney-piece in the dining-room of his house (which is still preserved) to testify to the health he had recovered. Anastasia Gray, benefiting by the waters, gave the stone chair affixed to the south wall of the King's Bath. At one time the wooden structure, called "the cross," in this bath was surrounded with crutches left there as mementos of the cure of those who had used them.

A more practical token of gratitude and one that was for a long time fashionable, was the placing of a massive brass or copper ring, inscribed with the name of the donor and fixed securely in the wall of the bath. As many as 213 of these gifts could formerly be counted in the different baths, but only 28 now remain. In the King's Bath there were 104; in the Queen's Bath 31; in the Cross Bath, 40; in the Hot Bath, 33; and since 1724 five have been presented. The alterations in the baths and the obliteration of many of the names explain the disappearance of so large a proportion of these votive offerings. On those that remain the Corporation caused the inscriptions to be renewed about forty years ago. One of these reads as follows: "I, John Revet, His Majesty's brazier, at 50 ye. of age, in ye present month of July, 1674, Received Cure of a True Palsie from Head to Foot on one side. Thanks be to God." Another has engraved upon it: "Barbara, Duchess of Cleveland, Anno Domini, 1674"; on the staple are the Royal Arms, with a bar sinister, surmounted by a coronet. A third is largely and handsomely ornamented, and has the legend: "Lydia White, Dawter of William White, citizen and draper of London, 1612." A fourth, given by Sir Thomas Delves, bart., of Doddington, in the county of Chester, is inscribed on one side: "Thomas Delves, B. By God's Marcy and Pumping here formerly ayded"; and on the other: "Against an Imposthume in his head caused this to be fixed, June the 13, 1693." There could be no better evidence of the large amount of suffering relieved and of the diseases cured by the hot springs of Bath than these emphatic tokens left by the grateful in past years. That they have lost none of their healing power every season's experience still attests, more particularly since the extensions made and the improvements introduced for using this bounteous gift of nature.

CHAPTER IV.

PLEASURE GARDENS AND SPAS.

MONG the amusements in vogue during the last half of the century were those supplied by the several pleasure gardens established on the confines of the city. The chief reason for their introduction was to induce visitors to remain through the summer months instead of decamping when the hot weather came, as the majority of them were wont to do. The new kind of recreation devised must, from the popularity it enjoyed, have answered its purpose, at least as long as it was regarded as a novelty. Scope was thus furnished for the beaux and belles to display their sartorial finery and graceful accomplishments, as well as for social intercourse, when the public rooms were not desirable trysting places. On the lawns and walks, amidst flower beds, shrubs, and statuary, there was ample room for the gay throng to disport themselves. The hooped skirts, plumed hats, and the powder and patches of the ladies were rivalled by the long-skirt coats, gorgeous waistcoats, tie wigs, silk stockings and buckled shoes of the gentlemen. The constant shifting and fresh grouping of the forms and colours made the scene kaleidoscopic. To see and be seen there was a duty not to be neglected by the devotees of fashion. Concert breakfasts, both public and private, were also given, followed by the stately minuet or cotillon on the green; even as early in the day as this was dancing a passion and delight to young and old.

Of the places provided for these entertainments the chief were Spring Gardens, the others being the Villa Gardens, King James's Palace, the Bagatelle, and, later, the Grosvenor Gardens. The contiguity of the first-named to the Baths and Pump Room gave them an unrivalled advantage. They were just across the river, "within 472 yards of the Pump Room as recently measured." The approach on the city side was not inviting. Visitors had to go down Boatstall Lane, pass the east gate to the bottom, on the left of which was a quay extending to Slippery Lane in Northgate Street. Here formerly stood a cross where the fish caught in the river was sold, and a ducking stool for the sousing of scolding women. In the middle of the century the stool was removed, and the fish market transferred elsewhere. At the north end was the ferry, the course of which is fairly indicated by the present Pulteney Bridge, and by it intercourse was maintained between the city and Bathwick and Hampton. People who did not care to traverse the slum on foot hired sedan chairs, which conveyed them to the ferry for sixpence. Once over the river they found the grounds delightful, and the scenery beautiful. The village of Bathwick consisted of a few houses adjoining the old church, and was almost lost amidst the wide expanse of meadows. The Gardens, therefore, were practically the country, and their airiness afforded a pleasant contrast to the narrow and stuffy streets of the city, while the grottos and arbours that adorned the grounds gave the subscribers cool retreats, where they could refresh themselves and partake of whatever luxuries the bill of fare offered.

Every Monday and Thursday a public breakfast took place to the sound of French horns and clarionets, vocal music by celebrated artistes being occasionally added. The tickets were 1s. 6d. each; for private breakfasts, without music, the charge was 1s. On Saturday evenings there was a public tea, also with French horns and clarionets, the admission to which was 1s., which entitled the bearer to tea or coffee. A speciality advertised were "Spring Gardens cakes and rolls," which were ready "every morning from early after nine," showing that our forefathers were wont "to brush with hasty steps the dew away" to enjoy these dainties. The subscription for walking in the Garden was 2s. 6d. for the season; non-subscribers were charged 6d., for which a ticket was given, entitling the holder "to anything at the bar of that value." One evening every week was devoted to illuminations and fireworks, similar to those at the London Vauxhall. From this it will be seen that the attractions were many, and the prices moderate. The Gardens, of course, did not escape the notice of Anstey, who describes a public breakfast given thereat by Lord Ragamuffin.

He said it would greatly our pleasure promote,
If we all for Spring Gardens set out in a boat;
I never as yet could his reason explain
Why we all sallied forth in the wind and the rain,
For sure such confusion was never yet known:
Here a cap and a hat, there a cardinal blown;
While his lordship, embroidered and powdered all o'er,
Was bowing and handing the ladies ashore.
How the misses did huddle, and scuddle, and run,
One would think to be wet must be very good fun;
For by waggling their tails, they all seemed to take pains
To moisten their pinions like ducks when it rains.
. . . .
So when we had wasted more bread at a breakfast
Than the poor of our parish have ate for this week past,
I saw all at once a prodigious great throng
Come bustling, and rustling, and jostling along,
For his lordship was pleased that the company now
To my Lady Bunbutter should curtsey and bow.

Some few insignificant folk went away,
Just to follow the employments and calls of the day:
But those who knew better their time how to spend,
The fiddling and dancing all chose to attend,
Miss Clunch and Sir Toby performed a cotillon,
Just the same as our Susan and Bob, the postillion.
All the while her mamma was expressing her joy
That her daughter the morning so well could employ.
Now why should the Muse, my dear mother, relate
The misfortunes that fall to the lot of the great?
As homeward we came—'tis with sorrow you'll hear
What a dreadful disaster attended the peer;
For whether some envious god had decreed
That a Naiad should long to ennoble her breed;
Or whether his Lordship was charmed to behold
His face in the stream, like Narcissus of old;
In handing old Lady Comefidget and daughter
This obsequious lord tumbled into the water;
But a nymph of the flood brought him safe to the boat,
And I left all the ladies a-cleaning his coat.

The Gardens continued one of the haunts of fashion until the building of Laura Place and adjacent streets, having encroached on the area, it was thought expedient to abandon the spot, more especially as fresh competition was threatened by the creation of Sydney Gardens.

The Villa Gardens are described as pleasantly situated at the extremity of the village of Bathwick, and as being accessible over the ferry opposite Walcot Parade. The domain was formerly the seat of Mr. James Ferry, from whom it was acquired about 1783. It was "neatly fitted up and laid out for the reception of company." Tea and coffee were always to be had, with horns and clarionets, and dinners and suppers at the shortest notice, the choicest wines being guaranteed. A visitor placed on record his impression of the place, for which he tried to obtain a rise at the expense of the rival Gardens. The house, he says, is a perfect palace, and the grounds are laid out with the greatest taste imaginable; the situation is likewise justly in its favour, being at a good distance from the river, on an eminence, and quite free from those damps which, if nearer, it would be subjected to. The evening was rather unfavourable, yet notwithstanding there was a numerous, genteel company, amongst whom were many persons of distinction. The music, which was well selected, was heard to great advantage, from the orchestra here being opposite the house. The illuminations were on a new plan, extremely elegant, and were heightened by a beautiful emblematic transparency. The fireworks were some of the best seen for a long time, particularly the naval engagement, which was a quite new thought, well designed and happily executed. Evidently there was no lack of enterprise in the management; but it did not save the Villa from gradual neglect and decay as a place of resort by the well-to-do.

King James's Palace came into existence before the Villa Gardens. It was so named from a tradition that King James II., after his abdication, took refuge within its walls. The house, however, was not built until half a century after this historical event. It is situated in Lyncombe Vale, and though some distance from the city, its sylvan seclusion and picturesque grounds drew a good many visitors. These generally took the ferry at South Parade, followed the footpath across the Dolemeads (skirting Ralph Allen's stone works), and came into the Carriage Road, thence to the rendezvous. The temptations to the public at this establishment were limited to refreshments and walking in the gardens, for which privilege the customary fee of 2s. 6d. was expected. Some years before the end of the century the property was sold, and ceasing to be a place of public resort it became known as Lyncombe House.

The Bagatelle (started in opposition to the above) stood at the junction of Lyncombe Valley road with the Carriage Road. Its amusements were akin to those of Bathwick Villa; but it had a special feature in a lake, the stream flowing through the valley having been intercepted to form it, with a pleasure boat on its bosom; a "luminary cascade" was another of its attractions. Its fate was the same as King James's Palace.

The Grosvenor Hotel and Gardens, "with an area of 24 acres square," were projected at the same time as Sydney Gardens, but the scheme, as before stated, was never fully carried out. The walls of the hotel were raised, but it remained incomplete for several years; it now forms the centre house of Grosvenor Place. Galas were given in the grounds, without, however, securing for the undertaking a permanent basis. The huge jawbones of a whale, which surmounted the eastern entrance gates, long remained a prominent landmark in the meadows.

Sydney Gardens Vauxhall (the original title) was a more ambitious undertaking than any of the others mentioned. In the rear of the handsome hotel erected, about 16 acres of land were reserved, and these were laid out with great taste and skill by Masters Harcourt. It was claimed for the Gardens that for beauty, situation, and variety they were not surpassed by any pleasure grounds of equal size in the Kingdom. Variety was undoubtedly a conspicuous element: there were a number of small delightful groves, pleasing vistas, some charming lawns, intersected by serpentine walks, while at every turn shady bowers, furnished with handsome seats, were provided—some canopied by nature, others by art. Waterfalls arrested the eye here and there.; stone and thatched pavilions and alcoves afforded agreeable seclusion or retreats for tea, coffee, and other refreshments. A sham castle, with several pieces of cannon, guarded bowling greens and swings. There was likewise a labyrinth, formed by hedge-bordered pathways, the principal one of which, with many intricate windings, led to a fine Merlin swing and a grotto of antique appearance. A spacious ride encircled the grounds. The amusements began early in the spring with public breakfasts, promenades, enlivened by music, and temporary illuminations. During the season there were four or five gala nights, when about 5,000 lamps were lighted and a "pompous" display of fireworks was given, to the great delight of the three or four thousand persons of fashion and consequence who on the average attended. The hotel at the entrance catered for the company and gave facilities for dancing and other games. The Gardens still exist, and though reduced in size and shorn of the artistic embellishments mentioned they are still beautiful, and admired for their walks and sylvan shade.

SPAS.

A curious manifestation of the century was the weakness shown for any spring of water that was credited with chalybeate qualities and certified for its medicinal virtue. Doctors readily took all such under their wing, and the afflicted, hearing of the cures effected, flocked to them for the relief of their ailments. Apparently the sufferers had more confidence in the cold water cure than in the nostrums of the faculty. "Throw physic to the dogs, I'll none of it," seems to have been the conviction on which sick people acted. They believed that a draught from an accredited fountain would make them healthy and strong at little cost. At the same time the newspapers of the period show us that quacks, with their testimonialised "cure-alls," flourished just as they do now, and will probably do to the "crack o' doom," for the gullibility of human nature appears to be beyond the power of education and culture to eradicate. They who preferred the rill to the "Balsam of Life," "Daffy's Elixir," "Rowley's British Herb Snuff," "Andrew's Nervous Pills," and the like chose the lesser of two evils, though it is a question whether the spas held in such repute were differentiated by their special virtues or their organic impurities. The failure of one and all is a sure proof either that the analyses made at the time were imperfect or that their remedial qualities were misunderstood or overrated. Yet such was the number of these "mineral fountains" in and around Bath, and so varied were their supposed constituents, that it was commonly asserted that there were few diseases incident to mankind but one or the other of them would cure!

The most noteworthy and most important was the Lyncombe Spa. Its discovery was due to Mr. Milsom, a cooper, who with others rented an old fishpond in the middle of the valley. A leak in the pond having occurred, he, in 1737, searched the ground at the head of it, and amidst some briars and willows found an open space, six feet long and three broad, which shook beneath his tread and looked like the spawn of toads, giving forth also a strong sulphurous smell. The slime having been removed several little springs came into view, bubbling up and emitting a black sand. A drop put into a glass of brandy made it of a purple hue, while three or four drops made it as black as ink. The water was proved by Mr. Palmer, surgeon, to be efficacious in gravel; and in the presence of the chief tradesmen of Bath and their wives he made a bowl of punch and pouring into it some of the water the liquor became of a blackish purple colour, and none dared to taste it until Mr. Palmer explained the cause, and the punch was then drunk with no little mirth and jollity. Dr. Hillary having analysed the water, and proved its health-giving properties, published in 1742 a treatise under the title of "An Inquiry into the contents and medicinal virtues of Lincomb Spaw Water," which he dedicated to the Earl of Chesterfield, who, he states in the preface, had experienced its virtues and had seen its effects in some remarkable cases. The water rose in two different springs, and though within a few inches of each other they were of different natures—one had a ferruginous smell, the other a sulphurous smell. The hard flint glasses used in drinking the water assumed in a few days a dark brown colour. The total yield was 1,800 gallons per diem, and as a result of the Doctor's analysis he was satisfied that "the water contained a much larger quantity of iron and sulphur than that of Harrogate, Tonbridge, or any of the hot springs of Bath, and that it was efficacious in the major part of chronical diseases," like that at Geronstere, in Germany, to which a resemblance was discovered. Confident that he had discovered an aqueous treasure he, in conjunction with Mr. Milsom, built a large and substantial edifice for the reception of patients expected at the Spa. Emblazoned above the fountain was the following inscription :—" The medicinal virtues of this water were first discovered by William Hillary, M.D." After all this preparation the spring disappeared, in consequence, it is supposed, of the ground having to be piled to get a good foundation for the house, which cost the unfortunate speculators £1,500. This was the building subsequently known as "King James's Palace." Later in the century "Lincomb Spaw" was still puffed, but the spring was on adjoining property, though it was represented to be the identical Hillary-Chesterfield fount.

The Lime-Kiln Spa, situated near a lime kiln at Swainswick, was long believed to be an effectual remedy for diabetes and other complaints. The owner made a cistern at the head of the spring, together with proper conveniences for drinking and bathing, and a dwelling house near, but an attempt was made to decoy the spring into lower land belonging to Sir Philip Parker Long's family, and it was so far successful that the water rose up in a meadow below, where a small portico was erected by the elder Wood for water drinkers. Wood deplores that "a spring that began to stand in comparison with St. Vincent's well, near Bristol, was by this division reduced to little or nothing."

The Bathford Spa was brought to light about the middle of the century, in the course of clearing an ash bed which grew about it. The water was observed to discolour every thing it touched, and samples having been sent for examination to Oxford and elsewhere with satisfactory results, people came to drink it for all kinds of maladies. A Bath physician bought the property, and built a square pavilion over the spring, which was henceforth known as a Spa. The neighbouring hamlet of Shockerwick was also famed for its St. Anthony's well.

Box likewise had its spa; a spring known as Frog's Well was reputed to contain the same salts as the hot waters of Bath, and its curative properties were correspondingly large. It was provided with a pavilion, and was for some time a favourite resort for invalids.

With these and other examples, people who owned property kept a keen look-out for any spring that might lie perdu, and likewise become a source of wealth to them as others had to their neighbours. Mr. Wicksteed, who was a noted seal engraver in his day, illustrates this eagerness. He "presents his most respectful compliments to the gentlemen of the Faculty in Bath, and, as he has great reason to believe that he has found a good cold chalybeate in his garden at Widcombe, which may be of service to Bath in general, and to many individuals in particular," asks for advice "how he may best secure the spring in the most convenient manner for the accommodation of such patients as they may think it proper to recommend it to." Wicksteed obtained so much encouragement that he soon after advertised his spring as a certain cure for consumption, the charge to patients being five guineas, and a cure promised in a month! The fee, however, was not to be paid if no benefit resulted. In other cases, such as " scurvy, relaxations, and obstructions," the drinking of the water was " by subscription, as usual." Wicksteed was joined by one Bowen, and in 1773 they were anxious to dispose of the property, which is stated to be commonly known as " Wicksteed's Machine or the Bagatelle," showing that it had been converted into a pleasure garden. It is now named Welton Lodge.

In addition to the Spas mentioned, there were sundry wells, the waters of which were held in high esteem. There were St. Winifred's well, which was visited more particularly in the spring of the year by people who drank the water, " some with sugar and some without " ; and the Carn well in Walcot Street, which was looked upon as impregnated with some fine minerals, causing the suffering to · go to it to fill their bottles and pitchers. On the east side of Beechen Cliff was " a remarkable spring which hath long been frequented by the ailing as well as by the sound." On the north side was the spring supplying the Cold Bath in Claverton Street, whose special qualities were attributed to the fact that it issued out of the ground at a place where the rays of the sun could never reach till after surmounting the equinox. In Frog Lane, on the north side of the town wall, was Frog's Well, which was much valued for its supposed beneficial effects in opthalmic cases. Road water was largely patronised, it being certified as a remedy in " chronical distempers " by Dr. Stephen Williams.

The ease with which people were beguiled by the healing reputation given to these spas and wells is strikingly exemplified by the Glastonbury Spring ; its popularity was such that it reduced materially the visitors to Bath for one or two seasons. In this instance a man named Matthew Chancellor, who was suffering from acute asthma, dreamed or fancied he dreamed that a spring of water, near the Abbey enclosure, was pointed out to him by a ghostly visitant, with instructions to use it for the cure of his complaint. He was to take a clean glass and drink it every Sunday morning seven times without any person seeing him. Chancellor did as directed, and was made whole. He published his story to the world, and Glastonbury was swarming with visitors, who believed what they had read and expected to find the like relief. As many as ten thousand strangers were quartered in and around the town at one time, and a depôt for the sale of the water was opened in London. A miraculous quality was imputed to the spring, because Chancellor's spiritual informant assured him that it rose " on holy ground, where a great many saints and martyrs had been buried!" The statement, if true, would have sufficed to condemn the use of the water in any but a credulous age. As it was, the faith of the people in these hydropathic remedies amounted to a mania, and as such it deserves to be added to the list of popular delusions.

CHAPTER V.

THE GAMBLING MANIA.

URING the period under review, gambling was carried to such lengths and involved so many tragedies as to make it a blot on the city's escutcheon. The vice was focussed here just as much as it is now at Monte Carlo. High play was all too common; the ruin it entailed, of which there were constant examples, failed to act as a deterrent; so strong was its fascination that the resistance of a wise resolve was often compelled to yield to its subtle power The memoirs and letters of the period contain, in fact, constant reference to the gains and losses (the latter especially) of individuals who came here, in many cases, more to enjoy the fraudulent games in vogue than to seek health from its healing springs or to find relaxation amidst its beautiful scenery. Gambling, no doubt, is and has always been, a universal weakness of mankind both in its savage and civilised state, and probably was never more rife among all classes in this country than at the present time. It is not confined to cards and dice, or the roulette table, as of old, but the craving for it obtains scope in betting on horses, football, cricket, and in speculating in stocks and shares, to which the multiplication of limited liability companies has given such a marvellous impetus. Whatever may be done by the Legislature, it is to be feared, judging by past experience, that the evil will go on practically unchecked, more particularly as neither religion nor education seems to exercise any restraining influence on the classes and masses infected by it. The greater the national prosperity the more gambling, like drunkenness, appears to flourish. "As a fellow feeling (it is said) makes us wondrous kind," it is quite possible that many may look with an indulgent eye on the shameless encouragement the Queen of the West for a long time gave to this pernicious pastime.

The century opens with a typical case. Installed as the director of amusements was one Captain Webster, whose chief qualification was his devotion to the gaming table, which cut short his career. A quarrel having arisen between him and another player over some stakes, a duel with swords took place in the Grove; Webster was run through the body by his antagonist, and his death followed immediately after. His mantle descended on Nash, who had been his coadjutor, and who, in his first season, is said to have won £800 at the green table, a sum that led him to conclude that for an impecunious adventurer there was no place like Bath. The solicitude he manifested for securing order and good behaviour among visitors was equalled by his assiduity in seeing that all due facilities were provided for gambling. With the taste then prevailing, there was nothing in his conduct in this respect that exposed him to censure. Indeed, it was a common thing for ladies of fashion, on falling into pecuniary difficulties, to open gambling-houses for the entertainment of all persons who played for fashionably high stakes, and for peeresses to insist on their privileges to keep "hells"—a privilege that was not abolished until 1744. According to the ideas of the time, Nash's action was venial. In his case it answered a double purpose: it enabled him to gratify his extravagant habits and to supply a want felt by visitors, to whom gambling was an indispensable excitement. Strange to say, he had the medical profession as an auxiliary. The local physicians were wont to prescribe play as a soothing relaxation to their patients! De Foe refers to the fact with his customary incisiveness. Speaking of Bath in 1722, he says: "The walks are behind the Abbey, spacious and well-shaded, planted all round with shops, filled with everything that contributes to pleasure. At the end is a noble room for gaming, from whence are hanging stairs to a pretty garden for everybody who pays for the time they stay to walk in. We have often wondered that the physicians of this place prescribe gaming to their patients in order to keep their minds free from business and thought, that the water on an undisturbed mind may have the greater effect, when, indeed, one cross throw at play must sour the man's blood more than ten glasses of water will sweeten it, especially for such great sums as they throw for every day at Bath." With this professional encouragement, it is no wonder that women were as passionately addicted to the vice as men.

Sir Richard Steele, who was not an infrequent visitor, enlarged on the rage for gambling he witnessed here in one of the papers contributed by him to the "Guardian" (No. 174), and published in 1713. He pays a visit to Harrison's table, and after describing in an ironical vein the devotion of the men to play, he turns his attention to the ladies. "I must own," he says, "that I received great pleasure in seeing my pretty countrywomen engaged in an amusement which puts them upon producing so many virtues, especially as they acquire such boldness as raises them near that lordly creature—man. Here they are taught such contempt of wealth as may dilate their minds and prevent many curtain lectures. Their natural tenderness is a weakness very easily unlearned, and I find my soul exalted when I see ladies sacrifice the fortunes of their children with as little concern as a Spartan or a Roman dame. In such a place as Bath is I might urge that the cast of a dice is indeed the properest exercise for the fair creatures to assist the waters, not to mention the opportunity it gives to display the well-turned arm and to scatter to advantage the rays of the diamond. But I am satisfied that the gamester ladies have surmounted the little vanities of showing their beauty, which they so far neglect as to throw their features into distortions and wear away their lilies and roses in tedious watching and restless lucubrations. I should rather observe that their chief passion is the emulation of manhood, which I am more inclined to believe, because, in spite of all slanders, their confidence in their virtue keeps them all night with the most dangerous creatures of our sex. It is an undoubted argument of their ease of conscience that they go directly from church to the gaming table, and so highly reverence play as to make it a great part of their exercise on Sundays."

Another writer, describing a visit to the room a few years later, remarks: "I could not help observing that several ladies, who had just been at prayers, seemed as earnest within an hour after at the diversion of 'O.E.,' a game at present much in vogue here. There were several parties beside at 'Whisk,' and numbers of lookers-on, who were generally thought by people of the house to encumber the room which others would better fill. Indeed, there were several who did not play at noon, but were making parties for the night." Had play been honourably conducted its victims would not have been so many, but this negative merit was too often lacking. Gangs of sharpers infested the gaming rooms, and, acting in concert, remorselessly fleeced the unwary. Nash was secretly in league with some of these, and shared with them their ill-gotten gains. Anecdotes are told of his kindly interposition to save youth from the folly of rash play, and their seniors from plunging. That he cared nothing for money, except in so far as it ministered to his pleasures, and spent it with a prodigal hand, is quite true, and it may be that he sincerely desired to save one and another from the certain ruin that he foresaw, particularly when he himself was on the crest of fortune's wave. Evidently, the victims were quite numerous enough to arouse now and again a compassionate feeling in the breast of the powerful M.C. It is recorded that not less than three persons fell a sacrifice in one day to this destructive passion. Two gentlemen fought a duel in which one was killed and the other desperately wounded, and a youth of great expectations at the same time ended his life by a pistol. Season after season, so deep was the play, that fortunes were lost, estates hampered, and families impoverished, as well as lives sacrificed. Lady Lechmere, who perished by her own hand, was said to have been driven to the desperate deed by her losses at cards. The most callous could hardly remain unmoved amidst such calamities. Nash had, no doubt, his qualms of conscience; besides, occasional displays of magnanimity helped to shield from disclosure his own participation in the plunder obtained. As Lord Chesterfield wittily retorted, when the Beau was complaining that he had lost £500, "I don't wonder at your losing money, but all the world wonders where you get it to lose."

Among other stories related of him to illustrate the better side of his nature are the following, though the sequel in each is somewhat equivocal:—The Duke of B. being chagrined at losing a considerable sum, pressed Nash to tie him for the future from playing deep. Accordingly the Beau gave his Grace a hundred guineas to forfeit ten thousand whenever he lost a sum to the same amount at play in one sitting. The Duke loved play to distraction, and soon after at hazard lost eight thousand guineas, and was going to throw for three thousand more, when Nash, catching hold of the dice box, entreated his Grace to reflect on the penalty if he lost. The Duke for that time desisted; but so strong was the furore of play upon him that soon after, losing a considerable sum at Newmarket, he was contented to pay the penalty. When the Earl of T——d was a youth he was passionately fond of play, and never better pleased than when he had Nash as an antagonist. Nash undertook to cure his Lordship, and engaged him in play for a very large stake. In proportion as the Peer lost the game he lost his temper too, and as he approached the gulf seemed still more eager for ruin. He lost his estate, some writings were put into the winner's possession, his very equipage was at last deposited and he lost that also. Nash rebuked the youth for his folly, and returned all, only stipulating that he should be paid £5,000 whenever he should think proper to make the demand. He never claimed the bond during his Lordship's life, but after his decease, the affairs of Nash being on the wane, he demanded the money of his Lordship's heirs, who paid it without any hesitation.

Nash was able to conceal for a long time his connection with "the black-sharping nest, the community's pest."; but the scandal of gaming in Bath and elsewhere became so notorious that Parliament tried by legislation to put a stop to the evil. In 1739 an Act against gaming was passed, which, after reciting that persons had, for many years past, carried on certain fraudulent games of chance with cards and dice, such games being the Ace of Hearts, Faro, Basset, and Hazard, and thereby defrauding several of his Majesty's subjects, it declares the said games to be lotteries, and imposes a fine of two hundred pounds on any person or persons setting up such games, two-thirds of which, if the offence was committed in the city of Bath, were to be given to the new Hospital there. Persons caught playing were to be fined fifty pounds. The Act was stringent enough; but it was reduced to a nullity directly it was passed by the invention of fresh games or the giving of new names to the old ones. "Passage" was the name of one of these, and a clause was forthwith added to the Horse Racing Act, declaring it illegal. When "Passage" went down, "Roly-Poly" (as Roulette was called) came up; and so the conflict went on between the virtue-designing Acts and the gambling propensities of the people. That the former became a dead letter will not be thought surprising when, in addition to the strength of the passion for play, it is remembered that not a few of the law-makers were among the law-breakers. These legislators having fulminated heavy pains and penalties in St. Stephen's against gamblers and their dupes, came to Bath and set the law at defiance, thus encouraging a vice they professed a desire to stamp out. Even some of the clergy patronised the tables, and one noted for his devotion thereto was styled "The Bishop" by the set in which he mixed.

The example set in the Royal circle was not one whit more consistent. In 1755 it was announced in the Press that his Majesty, the Duke of Cumberland, the Princess Amelia, the Earl of Northumberland, Sir John Bland, and John Offlay, Esq., played Hazard at Court, when the Duke won £4,000 and the Princess Amelia £1,000. The Court was, however, privileged, inasmuch as the Act passed in the twelfth year of George II.'s reign, while making it illegal to play games then known in any place, excepted "the Royal Palaces where his Majesty, his heirs or successors shall then reside."

The new game, as it then was, with which Nash was more particularly identified was called E.O. It was first played at Tunbridge Wells, where the Beau was a partner of the croupiers. Finding it very profitable, he determined to introduce it to Bath, first taking the precaution, it is said, of asking the opinion of several lawyers, who believed it to be in no way illegal. In consequence of this he (according to Fleming, the musician, who, under the pseudonym of "Timothy Ginnadrake," published some interesting reminiscences of Bath) wrote to Mrs. Hayes, who kept one of the Assembly Rooms at Bath, acquainting her with the profits attending such a scheme, and proposing to have a fourth share with her and Mr. Wiltshire, the proprietor of the other Room, for his authority and protection. Mr. Wiltshire and she returned him for answer that they would grant him a fifth share, which he consented to accept. Accordingly he made a journey to London, and bespoke two tables, one for each room, at the rate of fifteen pounds a table. The tables were no sooner set up at Bath than they were frequented by a greater concourse of gamesters than those at Tunbridge. Men of that infamous profession from every part of the kingdom, nay, even other parts of Europe, flocked here to feed on the ruin of each other's fortune and to fleece the inexperienced. How, under the stimulus given by Nash, play became the all-absorbing pursuit with both sexes is indicated in the following poem which appeared in the "Gentleman's Magazine":—

CUPID DEFEATED AT BATH.

Love told his mother t'other day,
　I'the Rooms he'd spend an hour,
See what they do, hear what they say,
　And try his utmost power.

He changed his form, he hid his wings,
　Clapt on a smart Toupee,
And was adorned with all those things
　A spritely beau should be.

On Celia first he fixed his eyes,
　She shunn'd his am'rous arts,
Impatient from the god she flies
　To Wiltshire's Ace of Hearts.

Fair Cloe next the youth addrest,
　With eloquence and skill,
But she her high disdain exprest,
　He could not play Quadrille.

Poems to Daphne he repeats,
　Translated out of Maro;
But wisely she prefers those cheats
　That deal to dupes at Faro.

Repuls'd by those, he talk'd to more,
　Yet talk'd alas! in vain;
From ten years old to full threescore,
　Their sole pursuit was gain.

The god enrag'd at their contempt
　Resum'd his native air,
Display'd his wings, his bow unbent,
　And thus bespoke the fair.

Adieu, he cry'd, ye giddy fair;
　I will no longer stay;
Those hearts deserve from me no care
　That are usurpt by play.

Hymen and I give up the field;
　Avarice now reigns alone;
Since cards and dice such joys can yield,
　Marriage and love are gone.

If Nash had looked after his own interests his share of the gains from the E.O. tables would have sustained his extravagant habits for a longer period, but, careless as he was, he trusted implicitly in the honour of his colleagues, and accepted without inquiry what they gave him. His portion, however, grew less each year, and at length he discovered that he had been cheated out of two thousand guineas, and on this basis he calculated that he had from first to last received twenty thousand pounds less than was his due. If this estimate was correct it shows how great must have been the profits of the confederates, as the Beau's share was only one-fifth of the whole. When the more stringent Anti-Gambling Act of 1745 checked these seminaries of vice, Nash found his prospects very cheerless, and hoped to brighten them by getting his quondam allies to disgorge, to do which they had not the least intention. They juggled with his demands, and, grown desperate, he commenced a law-suit against them. He was, of course non-suited, and not only did he lose the day; he lost credit with the public as well, for it was now known, what he had long endeavoured to conceal, that he himself was pecuniarily interested in the gaming tables, though before he had simply posed as the protector of them, and that he had part of the spoil, though complaining of having been defrauded of his just share. Notwithstanding this blackening revelation, his boundless assurance and bonhommie, the power that he wielded and the homage he exacted, together with the zeal he constantly showed in promoting works of charity, threw his peccadillos into the background; nor does it appear that his popularity suffered any serious diminution by the damaging disclosures of his abortive Chancery suit.

It is to the honour of the Bath Corporation that it was chiefly instrumental in getting the Act of 1745 passed, and the justices did their best to enforce its provisions. The "Bath Journal" of January 8th, 1749-50, gives the following paragraph:—"Last Tuesday Night, Charles Store, Esq., Mayor of this City, accompanied by several of the Corporation, and attended by proper Officers, went to a House near West-gate, kept by one R. Richards; for they had information that an E.O. Table was kept there. When they arriv'd they went up stairs to the Room where the table was, and found about sixteen Persons playing; the Gamesters, on seeing the Magistrates, were put into great confusion, and immdiately extinguish'd all the candles; two of them jump'd out of the window and made their escape, though one of them fell into the airy and was like to have been kill'd. The Magistrates prevented the escape of the others, procur'd some Lights, took down their names and sent 'em two at a time to Prison. They afterwards pulled down the Table, carried it into the Street, procur'd some Faggots, and burnt it; amidst a great Number of Spectators, who were very merry on the occasion." Despite this commendable energy and forcible example, the passion for play was too deeply rooted to be extirpated by law or its vigorous administration. The E.O. tables, though as illegal as the other forms of gambling of older date, were surreptitiously introduced and continued to lure dupes to their ruin. Indeed from the impunity with which the infamous calling was pursued for several years, the conclusion seems to be warranted either the authorities winked at the infraction of the law, or considered the eradication of the evil a hopeless task.

At length, the public indignation was aroused, to which the "Bath Chronicle" gave utterance on January 23rd, 1782. Referring to the E.O. tables abounding in London, it stated, as a notorious fact, that, "the keepers of these destructive traps for the unwary are wretches of the most abandoned principles and characters, noted swindlers and emancipated felons," and after this by no means flattering diagnosis of the fraternity, it adds: "With the most serious concern the respectable frequenters and inhabitants of Bath behold this place of sober amusement invaded in this daring manner, nor can they behold without indignation the various allurements thrown out to seduce the unthinking hopes of their families. To the magistrates, therefore, they look up to to suppress this scandalous and growing evil." The following week it returns to the charge, and that ignorance should not be pleaded to explain the supineness of the magistrates, it calls attention to "the notorious private E.O. table in Orchard Street," and declares that though justice has not yet overtaken the miscreants, the injured community retorts upon them its detestation and contempt. Possibly "the miscreants" deemed it prudent to "move on," as no proceedings are recorded against them. A year later another well-known haunt is presumably referred to in the subjoined notice:—"Private play is the bane of Bath. A well wisher to its public amusements cannot think it possible that anyone would dare to keep a Faro table and an E.O. table in the same house, in defiance of the respectable and spirited magistracy of this City; especially after the ignominy and punishments inflicted on the infamous practice in London." Thus boldly challenged the justices aroused themselves and vindicated the majesty of the law by burning a couple of E.O. tables. It is thus recorded in the "Chronicle" of April 9th, 1783:—"To the spirited zeal and activity of our Magistrates the public are indebted for the destruction of two E.O. tables, which were taken from two houses in this City and burnt in the Market-place on Saturday."

The game which proved so fascinating is thus described by Hoyle:—"An E.O. Table is circular in form, but of no exact dimensions, though in general about four feet in diameter. The extreme circumference is a kind of counter or depôt for the stakes, marked all round with the letters E and O, on which each adventurer places money according to his inclination. The interior part of the table consists first of a kind of gallery, or rolling place, for the ball, which, with the outward parts above, called depôt or counter, is stationary or fixed. The most interior part moves on axis or pivot, and is turned about with handles, whilst the ball is set in motion round the gallery. This part is generally divided into niches or insterstices, twenty of which are marked with the letter E, and the other twenty with the letter O. The lodging of the ball in any of the niches distinguished by those letters determines the wager. The proprietors of the tables have two bar holes, and are obliged to take all bets offered either for E or O; but if the ball falls into either of the bar holes, they win all the bets upon the opposite letter, and do not pay to that in which it falls, an advantage in the proportion of two to forty, or five per cent. in their favour."

The energetic action taken by the Justices failed again to keep out the blacklegs. Gambling flourished, stern as was the law and its probable enforcement. The "St. James's Chronicle" in 1787 regretted that there was no Bath Guide to inform the thoughtless gay young men who go thither that the first hour after their arrival, the depth of their purse is fathomed, and arts inconceivable formed by a junta of black-legs to strip them from top to toe, and that three out of five are constantly so stripped; when a young man therefore is invited to dine in private at one of the thick tabled houses, and not cautioned by the inviter about playing after dinner or supper, let him set the inviter down, whatever his name, rank, or condition may be, as one who has a design upon his purse. And as a proof of it if he declines play he will find no more invitation to an excellent dinner and the choicest wines.

The reproaches from far and near, in addition to the indignant protests from heads of families, aroused the magistrates to adopt more drastic measures. A man named Richard Westenell converted one of the houses in Alfred Street into a gambling saloon, where faro and other games could be played. The premises under his management had acquired a baneful reputation of which the Bench could not fail to be cognizant. Action was accordingly taken to suppress the nuisance. An information was laid before the Justices in April, 1787, summonses were issued, and Westenell, with one Richard Twycross, who was caught playing, were promptly arrested. At the hearing of the case there was no lack of witnesses—some of them having dearly bought the evidence they gave—to sustain the charge; and, after an adjournment for a week, the Bench fined Westenell £1,400 and Twycross £400. The decision gave rise to the following jeu d'esprit:—

> Well said, Master Mayor, by Jove, you don't joke,
> One thousand eight hundred brought down at a stroke:
> Proceed—and those black-legged locusts shall know
> E.O. shall be changed to a game of Oh! Oh!
>   Ha! Ha! Ha!

A fine of £1,800 in the eighteenth century would be equal to a fine of £5,000 at the present time. The penalty was thought to be excessive and was successfully appealed against at the Quarter Sessions. The "Chronicle" for April 25th, 1787, thus records the fact:—"Saturday last at the quarter sessions for this City, Mr. Westenell's fine for keeping the faro table, &c., in Alfred Street was reduced to £450 and treble costs, and Mr. Twycross's fine for playing was at the same time reduced to £100, which he paid with treble costs. At the adjourned sessions on Monday, a bill of indictment was found against the above house."

Although the penalty was mitigated, its exemplary character had a deterring effect. It kept the professional gamblers at a distance or prevented them from setting up establishments here for carrying on their nefarious trade. Doubtless the passion for play was as strong as ever, and found gratification among friends or in private houses where the gains and losses were quite large enough. The story of Midford Castle, if true, furnishes a striking illustration of this fact. It is said to owe its origin and peculiar shape (like the ace of clubs) to a large sum won at cards by Disney Roebuck at the end of the eighteenth century.

The difficulties in which people of quality often involved themselves by their devotion to play abound in domestic annals. It is recorded that the Duchess of D—— gave the following undertaking:—"Mr. D——ll having lent me £2,650, I do hereby promise to pay him £250 every three months at the usual quarter days, and continue to pay that sum quarterly to him or his heirs (allowing five per cent. interest, and five per cent. for the insurance of my life, per annum), until principal, interest and insurance shall be fully paid." In case of default, it was stipulated that the lender should acquaint the Duke with the transaction, and that in the event of her death, her Grace should leave a request that the money should be paid, the same having been lent "to relieve her from play debts, under a solemn promise that she will not play in future." Of another lady of title it is said that her debts were immense, that she had long lived torn by fear of the discovery of her liabilities, and tormented by the exigencies to which she had recourse to maintain concealment. How high were the stakes is shown by a remark in one of the letters of Sir Philip Francis, who states that he won £20,000 in one day at whist, which subsequent losses reduced to £12,000; and how general the love of play may be gathered from the correspondence of Mrs. Sheridan, who in writing to her husband had to confess that she had lost at one time 20 guineas, and at another 25 guineas at cards, excusing herself by the plea that she felt bound to conform to the custom of the society in which she was then moving. The broad acres of many an ancient house have by the same pursuit, followed, as is the rule, to infatuation, passed into the possession of strangers. True it is (as an old writer says) that the wealth which has been acquired by industry and hazard, and preserved for ages by prudence and forethought, is swept away on a sudden; and when a besieging sharper sits down before an estate the property is often transferred in less time than the writings can be drawn to secure the possession. The neglect of business, and the extravagance of a mind which has been taught to covet precarious possession, bring on premature destruction; though poverty may fetch a compass and go somewhat about, yet will it reach the gamester at last, and though his ruin be slow yet it is certain.

THE STORY OF SYLVIA S——.

Many more incidents might be given to show the disastrous effects on individuals and families through desperate devotion to play. To avoid an undue parade of illustrations, only two others will be given, owing to the pathetic interest attaching to the first, and the mystery partly enshrouding the second. The victim in the former was a young lady who dwelt in Bath in the early part of the century, and is known in contemporary records as Sylvia S——. Who were Sylvia's parents or where she was born cannot be stated, as they are unknown facts. She was, it is said, descended from one of the best families in the kingdom, and was as remarkable for her mental qualities as for her personal beauty, her letters and other writings, in prose and verse, as well as her conversational powers, indicating the gifts of mind with which she was endowed. In addition to these qualifications she was possessed of an ample fortune, of which she had the sole control before she attained the age of womanhood. Her parents must have died while she was quite young, as they are never found interfering with her life or imposing any restraint upon her actions. She had a sister, who passed away early, and her fortune was inherited by the survivor. Sylvia was thus possessed of means which would have enabled her to maintain a position in every way suitable to her birth and station. With this combination of attractions, it is needless to say how much her society was courted, or how many were the aspirants to her hand and heart. The number of her admirers was increased by her lavish generosity. She had no idea of the value of money, and in her thoughtlessness gave alike to the deserving and undeserving. Gay and good-natured, she fell an easy prey to designing knaves, of whom there were always a pretty good sprinkling in Bath; but her greatest misfortune was to become attached to Dame Lindsey, who owned the chief place of fashionable resort in the city. This woman, a selfish, dissolute creature, obtained a complete ascendancy over the lone girl, and whenever a person was wanting to make up a party for play at her house, Sylvia was at her command, although her docility not seldom emptied her purse.

A vague story likewise makes her figure in an unfortunate love affair. Among those who paid zealous court to her was a gentleman of position, but as imprudent and thoughtless as herself. She returned his affection, or what seemed such, and was looking forward to years of happiness in his society, when she found that he was heavily in debt. and that he had been arrested and was in gaol. With her usual kindness, she went to London determined to discharge the liabilities of her lover. Nash, who was in London and knew her well, endeavoured to dissuade her from her purpose, pointing out that it would be futile, as the scapegrace would, if freed now, speedily get into fresh difficulties. She, however, turned a deaf ear to the advice, the debts were discharged, and the prisoner was set at liberty ; but, as Nash predicted, he was soon again in difficulties, and died in gaol.

At Nash's suggestion Sylvia returned to Bath, and of her own inclination to the gaming table. She was as affable and as ready to oblige as ever, but her vivacity was broken by fits of melancholy, and she suffered in reputation through her close connection with Dame Lindsey and her allies. Wood, the architect, who likewise knew her, defends her character. "I could never (he says) by the strictest observations, perceive Sylvia to be tainted with any other vice than that of suffering herself to be decoyed to the gaming table, and at her own hazard playing for the amusement and advantage of others." He readily assented to a proposal that he should let part of his house (No. 24, Queen Square) to the young lady, who removed thither with an old and trusted servant in the summer of 1730. Though her losses and profuse generosity had made heavy inroads on her resources, she was ready to accept trifling marks of friendship to give her a pretence for making great returns. No sooner was she settled in her new abode than ladies of the highest distinction and of the most unblemished character were her constant visitors. Her levee looked, it is said, more like that of a first Minister of State than of a private young lady.

Far more pleasurable to Sylvia, in her less giddy mode of life, than these gay receptions were the walks she habitually took in the country. The beauties of the hills and vales around the city seemed to have an ever-increasing charm in her eyes ; and no wonder. The past could yield her no satisfaction ; the future was dreary and hopeless. The bitter alternative before her was—beggary or a uicide's grave. Nature, peaceful and lovely, was congenial companionship to her amidst a desolation, the secret of which was lodged in her breast alone. It was a balm and a solace, bringing those moments of forgetfulness so sweet to the forsaken and the sorrow-stricken. Meanwhile life was becoming more and more a burden to her, and she resolved to put an end to it. She delayed the fatal remedy till she had parted with her last farthing, as well as with everything whereby she could, with the least show of credit, raise a penny, and till she had performed a promise to take care of Mr. Wood's family while the other part was absent with him on a London journey.

During this interval a master workman lodged in the house for the better security of the inmates. Wood was detained in London longer than he expected, and at the close of the day appointed for his return Sylvia expressed some uneasiness at the disappointment she was likely to meet with in not surrendering up the trust she had taken upon her. She then sat at the dining-room window, and the following lines were subsequently found written with a diamond on one of the panes :—

O death! thou pleasing end to human woe,
Thou cure for life, thou greatest good below;
Still may'st thou fly the coward and the slave,
And thy soft slumbers only bless the brave.

It was noted that she was more than usually cheerful that evening—a common result in these cases when resolve supersedes hesitation—and, after fondling two of the children for some time, she went into the nursery to take leave of another child as it lay sleeping in the cradle, its innocent looks eliciting warm encomiums from her. She then retired to bed, but speedily rose, dressed herself in clean linen and white garments of every kind, "like one that was going to church to be made a joyful bride, her gown being pinned over her breast just as a neat nurse pins the swaddling clothes of an innocent babe." A pink silk girdle was placed round her neck, which was lengthened by another made of gold thread. The end of the former was tied with a noose, and that of the latter with three knots at a small distance from one another. A dressing stool was placed near a closet door to stand upon. After these preparations had been made, she sat down and read part of Sir John Harington's translation of "Orlando Furioso." The book lay open at the story of Olympia, who by the perfidy and ingratitude of her bosom friend was ruined and left to the mercy of the world. Next, mounting the stool, she threw the girdle over the top of the door, and, turning her back to it, shut and locked it; thrusting the stool aside, she became suspended, but her weight broke the girdle, and she fell down with great violence. The noise awoke the workman in the house, but hearing Sylvia walking about the room, as was her custom when she wanted sleep, he took no further notice of it. In the meantime, the unhappy lady had exchanged the silken girdle for one that was stronger and made of silver thread. She made a second attempt, which was successful; and the next day, when the room was forced open, she was found "dead, cold and stiff."

Such was the sad end of poor Sylvia. Gifted with qualities that would, under proper guidance, have made her an estimable member of society—an Elisabeth Fry or a Florence Nightingale—her sympathetic nature and impulsive generosity, her inability to say "no" when prudence suggested, proved her ruin, instead of being the foundation of a reputation for practical benevolence. A cruel fate doomed her to mix with the most depraved elements of society—the gambling, foul-mouthed debauchees who made Bath their rendezvous. Living so much in this social sewer, the marvel is that she maintained her virtue and the fascinating manner that all who came in contact with her found so attractive. She had, we are told, a strong belief in the immortality of the soul, but of the vital principles of religion that restrain and sustain she evidently knew nothing. Without these she drifted helplessly down the stream on which her frail craft was launched, until the burden of despair sank her for ever beneath its turbid waters.

When her death became known a Coroner's inquest was held on the body, and a verdict of temporary insanity was returned by the jury. Her remains were buried the next night in her father's grave in the Abbey church. the expense of the funeral being borne by a lady friend of the deceased. The register of the Abbey does not record the interment, otherwise the surname of Sylvia could have been given with certainty. She had set at nought the Divine prohibition against self-slaughter, and admission to the parish register was denied her. It may be that her name was Sidney; and from the allusions to the high position of her family it is probable in that case that she was an offshoot of the Sidneys of Penshurst, like Sir Philip Sidney, Algernon Sidney, and other celebrities of that historic house. If this supposition were confirmed it would give additional interest to her story.

Five months elapsed before anyone would deal with Sylvia's effects. At length Nash, the faithful servant of the deceased, was made administratrix, and the goods were then sold. Such was the eagerness to obtain souvenirs that the prices realised sufficed to pay all the creditors in full and to leave an overplus for the nearest relation. Wood very justly condemns the heartlessness of Sylvia's society friends, who, when they knew her fortune was ebbing, drew her into play to win her money, and accepted whatever was offered by her generous hand; but he himself is not free from reproach, his conduct exhibiting a meanness and littleness which contrast strikingly with his greatness as an architect. At the time of her death, Sylvia, he says, owed him "two and fifty pounds, three shilling and fourpence for rent," and in consequence "he seized all her papers and other effects on Monday the 13th September, 1731." He also complains that the surplus from the sale was not handed to him in addition to his rent, "as a consideration towards the damages I sustained on the score of Sylvia's untimely death!"

Stranger still is the sort of Walpurgis Night that he passed after receiving news of Sylvia's fate. It reached him at Frocksfield at sunset on his way to Bath. "The surprise (he writes) was so great that every bush I galloped by looked like an infernal spirit; every stone and clod of dirt that lay in the road appeared like a hobgoblin; and stone walls resembled nothing but swarms of dreadful spectres. The rustling of the trees and the sound of the horse's feet filled my ears with nothing but the groans and howlings of people in the utmost distress; and if the poet that described the journey of Ulysses to the lower regions of the dead had, in imagination, seen those objects and heard the sounds which in riding near thirty miles at the close of the day were perpetually presenting themselves to my mind and ears, his account of a world of darkness, inhabited by phantoms of the dead, might have received such improvements as would have filled every reader with horror and surprise." That a man under such circumstances should have been the subject of these hallucinations is by no means creditable to his self-possession; that he should have proclaimed his weakness to the world savours of imbecility.

Reading between the lines of Sylvia's life the thought is suggested whether Beau Nash may not have been quite as much her evil genius as Dame Lindsey. Ardent and impressionable, she may have been early captivated by him. He was in the prime of life, dashing and jovial, the most conspicuous figure in fashionable society, the members of which paid deferential attention to his edicts. With these qualities that dazzle, he united the arts that beguile, and as women (according to Byron) "like moths, are caught by glare," it is easy to conceive how attractive would be the personal attributes of the Beau to an impulsive young lady, sole mistress of herself and fortune. Neither were the pressing embarrassments, common to those who have to seek a living by their wits, unknown to him. When the two first met his income was very precarious. He had not then matured the scheme by which he eventually secured a handsome revenue from the gaming table. His necessities made him selfish and unscrupulous; he toyed with the young lady's passion while directly or indirectly he drained her purse. It is true that she is said to have squandered her money on a nameless lover, whose vices gave him the obscurity of a gaol; but Nash was an adept in inventing stories to conceal the sources of his gains. It is true, also, that he advised Sylvia to leave Dame Lindsey's and take shelter with Wood in Queen Square; but that was when her fortune had been dissipated, as is shown by the straits she was in during the very few months she lived under Wood's roof. Certain it is that she was in London at the same time as Nash, and it is significant that she changed the patronymic of her faithful attendant to that of Nash, as if to find an excuse for having constantly on her lips a name around which her tenderest feelings had been twined. If Nash really betrayed and abandoned her, then her case would be the counterpart of Olympia's in the "Orlando," and would explain why she left the pathetic episode in Ariosto's story to proclaim the cruel wrongs of which she herself had been the victim. The inference thus drawn of the relations between Nash and Sylvia is obviously a conjecture and nothing more; its acceptance or rejection is left to the reader's judgment.

A FATAL DUEL.

The other occurrence now to be narrated is connected with one of several duels that took place on Claverton Down, when the Grove had ceased to be the trysting spot for such encounters. The antagonists in this case were Viscount du Barré and Count Rice, both hailing from France. The former was the only son of Jean Baptiste, Count de Cérés; the latter was an Irish Jacobite, whose grandfather accompanied James II. to the Continent, and who had borne arms in the French service. Both were fast friends, and both were addicted to play. It is assumed that their chief object in coming to Bath was to find "pigeons" to pluck, and the more readily to favour their operations they became tenants of No. 8, Royal Crescent, where they gave grand parties and played for high stakes. The house was graced by the presence of the Viscountess du Barré and her sister, ladies of great beauty and accomplishments. One evening the sum of £650 was won by them from Colonel Champion, and there being some arrears due to Rice, he claimed the whole in payment. Du Barré objected, and taunted his companion with having decoyed him to Bath by false expectations. Incensed by the imputation, Rice threw down the glove, which was accepted, and, with the fierceness of implacable hate, they resolved that the meeting should take place forthwith. Sallying forth in the short hours of a gloomy November morning, they went to the Three Tuns, in Stall Street, with their seconds, and, hiring a coach, proceeded in sullen silence to Claverton Down. By the first break of day the preliminaries were settled and the ground marked out. Each combatant was armed with a brace of pistols and a sword. On the signal being given, Du Barré fired first, and lodged a ball in Rice's thigh, which penetrated as far as the bone. Rice then took aim and wounded his opponent in the breast. He went back two or three paces, and then advancing the pistols were again presented by each; both flashed in the pan, but only one was discharged. Throwing both weapons away, the combatants drew their swords; but, as they advanced to meet each other, Du Barré fell, exclaiming as he did so, "Je vous demande ma vie," to which Rice replied, "Je vous la donne." In a second or two the Viscount fell back and expired. Count Rice was conveyed to Bath with difficulty, owing to the dangerous nature of his wound. From some unaccountable negligence the body of the Viscount was left exposed the whole day on the Down, but was subsequently removed to Bathampton Churchyard, where, after the inquest, it was interred. A gravestone placed over the grave is inscribed. "Here rest the remains of John Baptiste du Barré. Obiit 18th Nov., 1778." The post boy found in Du Barré's overcoat pocket twenty diamond waistcoat buttons, which he attempted to conceal, but revealed them in his cups, and they were restored to the Viscountess. The contents of a letter also found threw no light on the tragic event, though it related to money transactions. On his recovery, Count Rice was tried for manslaughter, under the Coroner's warrant, at Taunton, but was acquitted. He subsequently went to Spain, and died there in 1809. A stone marks, or did mark, the spot on the Down where the duel was fought. A singular coincidence occurred in connection with it. On the Saturday following, Henderson was playing Falstaff at the theatre, and when he came to the words: "What is honour? A word. What is that word honour? Air. A trim reckoning! Who hath it? He that died on Wednesday!" the sensation through the house was most profound, Wednesday having been the day on which the duel occurred.

It should be added that no authoritative statement was ever made as to the real cause of the quarrel between the two noblemen. The secret was known only to Count

Rice, and he carried it with him to his grave. Of the two or three explanations current at the time, the one above appears the most plausible but whether exact or not it may be taken for granted that gambling was at the bottom of the dispute. The warning this, like so many other similar stories, conveys is not likely to have any influence on those given to play. A resolve to desist at a particular time, or under certain conditions mentally laid down, is scarcely ever kept, even when the conditions have been fulfilled. The habit grows into a passion which the will is more and more powerless to restrain. It is only when ruin—the usual penalty for this infatuation—mocks its dupe that the folly of the pursuit is fully realised, and that vain regrets add an additional canker to the misery from which the only escape is death. Far better would it be if players would do as the young lady did upon whom the following lines were written extempore by one who saw her play at E.O. :—

With manners gentle as the stream
  From whence her blood did flow,
A nymph sat down, with graceful mien,
  The circling ball to throw.

No anxious cares disturb'd her breast,
  Nor sordid gain her soul possess'd;
Unmov'd she sat, if Fortune frown'd
  Or flatter'd as the ball went round.

To E or O she seem'd inclin'd,
  As Fancy painted to her mind;
Success rewards the prudent Fair,
  And rais'd a pile of silver there.

The Goddess* blind, with envy seiz'd
  To see her favours thus despis'd,
Snatch'd from her hand the silver store,
  The lady smil'd—and play'd no more.

* Fortune.

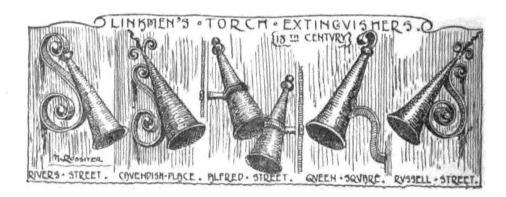

LINKMEN'S · TORCH · EXTINGUISHERS.
18TH CENTURY.

RIVERS · STREET.   CAVENDISH · PLACE.   ALFRED · STREET.   QUEEN · SQUARE.   RUSSELL · STREET.

CHAPTER VI.

MUNICIPAL AND POLITICAL. ·

N the middle of the century the city contained seven-and-twenty streets (thirteen new), six terrace walks, seventeen lanes and nine gates, four throngs, four alleys, two bridges, six courts, and ten open areas; with five hot baths, one cold bath, eleven conduits, six hospitals, three churches, three chapels, three meeting-houses, a court of justice, a gaol, two assembly rooms, two schools, and two poorhouses, besides the remains of three other churches and two chapels—monuments of indifference and neglect. Civic affairs were conducted on the lines laid down by the charter of Queen Elizabeth, which provided that the Corporation should consist of a mayor, a number of aldermen—not less than four and not exceeding ten—and twenty chief citizens, to be called common councillors, by whom were to be elected a recorder, common clerk, chamberlain, or receiver, constables, and other inferior officers, with two sergeants of mace. Watchmen and beadles were also to be appointed annually for each of the parishes in the city, with directions as to how often and in what manner they were to go their rounds. In addition, one or more of the inhabitants of the parish had to keep watch and ward between the hours of nine in the evening and seven in the morning to see that the Dogberries performed their duty—not an exhilarating task, certainly not in wintry weather, when a comfortable bed had to be exchanged for a prowl after "jarvies" through a labyrinth of streets, dark almost as Erebus, and cheerless as London in a fog. The cost of the system was defrayed by rates levied upon each parish. The lighting of the streets, lanes, alleys, and other public places was left to the caprice of the inhabitants; every householder, according to a by-law, was enjoined, from the 14th September to the 25th March, to hang out candles or lights in lanthorns, on the outside of his house, "as it shall grow dark, until 12 o'clock at night, upon pain of forfeiting two shillings for every default." In like manner the duty of keeping the streets in repair devolved upon the citizens; every owner or occupier of any house or land adjoining any thoroughfare was required, on receiving official notice, "well and sufficiently to pitch or pave" one half of the street in front of his premises within ten days after notice, or forfeit the sum of ten shillings for each perch left unrepaired.* Furthermore, three times a week the householders were expected "to sweep and cleanse, or cause to be swept and cleansed," all the streets and lanes before their respective houses; but the by-laws, wholesome in theory, were not systematically enforced. The rulers connived at their infraction because it relieved them from their

---

* Householders were not relieved from this duty until the passing of the Bath Act, 1851.

responsibilities as citizens. Thus the public interests were sacrificed to private parsimony. It was the same with the chairmen. There were laws for licensing, regulating, and governing them :—They were only licensed for one year; their chairs were to have a mark or figure for identification; they were confined to certain appointed stands; and their fares were duly fixed, but with all these precautions the authorities were constantly set at defiance by this turbulent and insolent body. Strong in numbers and firmly united for the protection of their interests or their extortions, neither the law nor its administrators had any terrors for them; they did pretty much as they liked. Before the wearing of swords was prohibited it was not at all uncommon for the chairmen to provoke gentlemen to draw their swords and then to make a ferocious attack with their chair-poles in so-called self-defence. Though they banded together for mutual protection in emergencies, there were constant quarrels among themselves when waiting for employment at the public assemblies, causing confusion and alarming ladies. On one occasion a noble Duke, with several other distinguished personages, having stayed after a public ball was over, found at the door a number of chairmen eager for employ; but the night being bright and warm the company declined to make use of the chairs, resolving to walk to their respective lodgings. When near the Churchyard they were pelted with mud " by some invisible hands," the owners of which were as a class well known, but though Mr. Nash offered two guineas and the Corporation twenty guineas for the discovery f the offenders, no betrayal was made, notwithstanding, too, that a free pardon and the above rewards were promised to any one informing against his accomplices. From an incident already given it will have been gathered that sobriety was a virtue not common among these men. In the autumn of 1748, "a nobleman of the first rank" publicly caned his chairmen upon the Grand Parade for their inability, through intoxication, so much as to hold up the head of the chair while his grace was quitting it. The thrashing of the fuddled fellows by a peer of the realm in a frenzy of indignation must have been a sight worth seeing, albeit the victims of his wrath may, from their confused mental condition, have been uncertain whether it was a playful or hostile manifestation. In course of time greater respect for law and authority was diffused among the chairmen, who it was usual to swear in then as special constables. They were about 200 strong.

The foregoing details will give some idea of civic management in the " good old times." How the Corporation was elected and how it discharged the fiduciary duties devolving upon it will now be explained. Its existence was maintained on what is now known as the co-optive principle; when a vacancy occurred the remainder appointed any person they liked to fill it. It was, therefore, a close or self-elected body, on which public opinion could, as a rule, exercise no influence. Acting in secret, and having the control of the public charities, it was guilty of gross and systematic malversation. All the estates of St. John's Hospital, as well as the extensive property in lands and messuages, tenements and rents in and near Bath, which Edward the Sixth granted for the erection of alms-houses and a Grammar School, were confided to it by the charter, and excellent rules were prescribed. But neither law nor honour was strong enough to restrain the rapacity of these trustees. No adequate provision was made for boarding and lodging poor people, and after some sort of Grammar School had been opened it was left without a master and its doors were closed. The wealth intended for these benevolent objects was diverted for the benefit of the members of the Corporation and their relatives. From the way in which the charities were plundered for a century and a half it would almost seem as if their existence, or the intentions of the Royal founder, had been forgotten (although there were not wanting ruins to keep them in remembrance), and that the members cherished the delusion that they were simply dealing with perquisites attached to their office. If at any time one less callous than the rest had qualms of conscience he was soothed by the assurance that it was alright, as it was what their predecessors had done. So the frauds went on from year to year.

But the day of reckoning came at last. A transaction of more than usual turpitude took place in connection with St. John's Hospital. The master, the Rev. William Clement, granted a lease of all the hospital property to his son at a rental of £130 a year. Conscious of their own evil doings, the Corporation, as trustees of the hospital, durst not interfere lest they should be exposed to recriminations. Their passive attitude did not commend itself to the successor of Clement, the Rev. John Chapman; he filed a bill in Chancery against the lessee, and under an award of Sir John Trevor, Master of the Rolls, in 1717, the lease was set aside, and the income from the property apportioned in a more equitable manner. Both the hospital and chapel attached to it were at the same time rebuilt. As an additional security it was decreed that the Lord Chancellor, the Lord Keeper, the Master of the Rolls, and the Lord Bishop of Bath and Wells, or any two of them, should from time to time respectively for ever be visitors of the hospital. Of the warning thus given the Corporation took no heed; they did not attempt to mend their ways. True it was that a Grammar School had been re-started in a mean way, the body of St. Mary's Church, which stood near the North gate, being devoted to its use, while the tower above was the city prison, or a den of thieves. With this economical arrangement the guardians were content; but not so Mr. Walter Robinson, who was appointed to the mastership in 1730. Discovering how shamelessly the charity had been abused, he also sought the aid of the Court of Chancery, and as a result of the writ issued against the Corporation, it was found that the Grammar School had been starved, that the " able schoolmaster " received only £20 a year and sometimes only £10, and that, too, though there had been a great increase in the value of the lands and tenements given by King Edward; further, that no provision had been made for the relief and support of any poor whatever; on the contrary, that the Corporation had notoriously mismanaged, misconverted, misgoverned, and misapplied the revenues. These iniquities being fully established, the Court ordered the Corporation to place the charities forthwith under the care of trustees, who were required to carry into execution faithfully the intentions of the Royal donor. In consequence the Grammar School was provided with a commodious building in Broad Street, combining a residence for the master's family and convenience for boarders, in addition to the free scholars. The salary of the master was increased to forty guineas per annum, and the Corporation were ordered to pay him fifty pounds to defray the legal expenses he had incurred. All the accounts were to be entered in a book by a duly qualified official, and a duplicate of the said book was to be kept in the library of the Abbey Church. St. Catherine's Hospital was likewise re-opened for the lodging and maintenance of ten poor folk; but in both cases the work of reform was hampered by the fact that the property had been leased at such ridiculously low rents to the civic worthies or their friends that the gross income receivable was very small. When the leases had expired a larger measure of benefit would, it was expected, be dispensed through the charities dealt with. By these changes the more glaring abuses were removed; but there was much that called for rectification at a later period, particularly in regard to the management of the resources of St. John's Hospital, the mastership of which remained in the hands of the Chapmans throughout the century.

THE WATER SUPPLY AND PUBLIC CONDUITS.

The improvement of the thoroughfares of the city involved the sacrifice of much that was interesting both from an antiquarian and architectural point of view. Not the least of these losses were the fountains or conduits, which gave, in addition to that conveyed to private houses, a supply of water adequate for the wants of the population. It would, indeed, have been surprising had there been any deficiency. The hills around yielded copious springs, one or two of which the Romans, with their customary attention to sanitation, utilised for their public buildings and were doubtless made available for general use. At a later period the monastic authorities laid down mains, and to them is due the provision of those large and tastefully built conduits to which all had access. Captain Chapman, writing in the last quarter of the seventeenth century, speaks of the "crystal springs gushing out of the hills," from two of which, on the north and south side, "the city, by pipes of lead, were not only plentifully served with common conduits, but not a few of the houses were supplied within their own doors." The abundance of springs enabled the principal ground landlords, when the city was rapidly growing, to supply without much difficulty the needs of their tenants, and to obtain an additional income for themselves from the waterworks thus created. Supplementing as these did the resources of the Corporation, explains why the Health of Towns Commissioners, when they visited this district in 1844, were able to report that "the city of Bath affords an instance of a town generally well supplied with water without any legislative provision for the purpose." It is a singular commentary on this favourable opinion that the Town Council within a couple of years began to obtain Acts of Parliament for improving the supply; that since then they have spent nearly a quarter of a million on water works, or in buying up the private companies and in extending their mains beyond the borough limits. And the cry is still for more storage space.

Of the old conduits a description has been happily left by Wood. The principal was St. Mary's, which was a beautiful quadrangular edifice, standing in the middle of High Street. It was of the Doric order, covered with a cimasium roof, which, after passing through a ball, terminated in a point; every corner of the structure was adorned with a pinnacle composed of four stones, the last of which bore a globe; the water issued out of each side of the body of the building. To this fountain the citizens were wont to make their grand processions, the most noteworthy of which was on the Coronation Day of Charles II., 1661, when "Mrs. Mayoress, attended by all the chief matrons of the city, and preceded by above four hundred young virgins, going two and two, bearing aloft in their hands gilded crowns and garlands, decked with the rarest and choicest flowers of the season, went to the conduit to drink the King's health with the claret that then run from it."

The architecture of St. Michael's conduit, which faced the south front of the church, was still more imposing. Its base made a perfect cube, and elevated its tower to a considerable height. Four Ionic pilasters composed the structure; these standing upon a pedestal, the base and body of which were turned into one large cimasium, convex below and concave above; the entablature was surmounted by five plinths or steps; and these bearing a proper pedestal, a double plinth upon that, sustaining an ornament in the shape of an hour glass, crowned the whole tower. Four niches adorned the four fronts of the structure between the pilasters, and four shields were carved on the four faces of the body of the upper pedestal. The water issued out of every side of the base in the centre of a semi-circular arch, intended to represent a rainbow. The conduit of St. Peter and St. Paul was in the shape of a small High Cross, and stood before the front of the Guildhall. Broad Street conduit was an ornamental building, placed in the middle of the street. Cornwell spring in Walcot Street was surmounted by a High Cross or tower. Stalls conduit made the termination of the central lines of Cheap Street, Westgate Street, and Stall Street; St. James's conduit faced the Southgate, both being artistic erections. The substantial and ornamental character of all these structures shows that even in minor matters mere utility was scorned. It was considered that buildings dedicated to the public service were deserving of being made beautiful as well as useful, something to embellish the city and to gratify the eye and taste. No doubt the cost of keeping such grand conduits in repair was felt to be a burden, and being shirked, caused dilapidation, which furnished an argument for their removal, in addition to that supplied on the score of the obstruction to traffic. No sentimental regard for their antiquity seems to have urged a plea for the preservation of any one of them. All were doomed to perish, and so not one of these links with a remote past existed beyond the middle of the eighteenth century. The enlightened destroyers were content with a single tap for water being affixed to some wall or house near where the towering conduit stood. Wood complains that the supply was not adequate, and even urged the Corporation to collect the springs into a large reservoir on the highest part of the town so as to meet more effectually the public requirements. It is another evidence of the great architect's practical genius and foresight. A century had to elapse before the Corporation could boast of having a reservoir of even moderate size.

POLITICAL.

One duty the Corporation certainly discharged with credit. Upon it devolved the selection and election of the members to represent the city in Parliament. In this important matter the citizens outside had no more voice than in the smallest detail of the civic administration. They had to be content with learning the names of the two gentlemen chosen at a General Election, or the one appointed when an extraordinary vacancy occurred. The authority of the Chamber was absolute, and could not be called in question by the constituency for whom the representatives were provided. It had simply to rest and be thankful. Though the position, according to modern notions was anomalous, the trust reposed in the authorities was never abused. Men of repute in the city, or county magnates without, as well as others of eminence in the State, were sought out for the office. Looking through the roll of members, it will be seen that, mingled with the local names of Gay, Chapman, Moysey, Trotman, and the like, are those of Codrington, Blathwayt, Langton, Popham, Horner, Thynne, etc.—all persons of position and substance. The avenue to Parliament thus opened entailed little trouble or expense on the aspirants. There were no published addresses to the electors, no canvassing, no noisy public meetings, no agents, and no "itching palms" to be soothed. An interview and an interchange of opinion with the members of the Corporation sufficed to transact the whole business. The only expense was a dinner to the Council and largesse to the crowd on the chairing day. Very different this to some modern elections; one that occurred before the passing of the Corrupt Practices Act cost the candidates something like ten thousand pounds (the price of an old pocket borough), in addition to an exhaustive expenditure of voice and energy, and a generous bespattering of abuse. Under the old regime the members chosen had their pockets so lightly taxed that they could very well be liberal contributors to public objects, if so disposed, as, in particular, was Field-Marshal Wade. Of him and one or two other political notables who were identified with Bath in the period under review some notice will now be taken.

Wade was the third son of William Wade, a major of dragoons in Cromwell's troops. He entered the Army in his seventeenth year, and saw considerable service under William III. and Marlborough. His abilities must have been conspicuous as his promotion was rapid. In 1715, when he is first heard of in Bath, he had attained the rank of Major-General, and in that year he was returned member for Hindon, Wilts, his political views being decidedly Hanoverian. A prominent part was taken by him in suppressing the Stuart risings, in one of which, that of the above year, this city was seriously implicated. The Western counties were thought to be ripe for revolt under the leadership of the Duke of Ormond, and among the precautions taken by the Government were the strengthening of the Royal forces in Bristol to prevent the seizure of the castle there; and the despatch of Rich's dragoons and another regiment, under the command of Wade, to Bath, which was the rendezvous of the conspirators and an arsenal for warlike stores. Through Ralph Allen, who was then a clerk in the Post Office, the General, it is said, obtained information that enabled him to seize eleven chests of firearms, a hogshead of basket-hilted swords, another of cartouches, three pieces of cannon, one mortar, and moulds to cast cannon. A number of horses were impounded, and several suspected persons were arrested and conveyed to London. Allen married Miss Earl, a natural daughter of Wade, and this fact is thought to confirm the story of the assistance he rendered the General in checking the conspiracy. A recrudescence of the Stuart plots in 1719 brought Wade back from the Continent, where he had been campaigning, to meet the danger. Bath was again implicated. Carte, who was minister at the Abbey, was a zealous Jacobite, and succeeded George Kelly as agent between Atterbury and the Pretender. He journeyed to and from the Continent in carrying out his intrigues, of which the Government, through its spies, was kept well informed. As it held the threads of the conspiracy as well as Carte it did not care to arrest him; while he was the chief emissary of the Jacobites it knew where to look for particulars of contemplated plots, and to watch the pivot on which these were likely to turn. Whether he was in danger of capture or not, Carte escaped by jumping from a window "in full canonicals," and reached a place of refuge. It is said that he died of a broken heart when he found that Prince Charles, for the sake of Clementina Walkingshaw, had broken with the English Jacobites. The documents collected by him for historical works he bequeathed to the Bodleian Library, Oxford.

The renown of Wade, and the influence of Allen, commended him to the notice of the Corporation, by whom he was elected one of the representatives of the city in 1722, having John Codrington, of Doddington, as his colleague. He was re-elected in 1727 with Robert Gay; in 1734 again with Codrington; in 1741 with Philip Bennet; and in 1747 with Robert Henley, Recorder of the city. His popularity in Bath was fostered by his generosity and genial manners, as well as by his martial prowess. The portraits of the members of the Corporation were painted at his expense after his second election, and hung in the Guildhall; a few of them still exist. At the request of his constituents the General's portrait, full length, was painted by Van Diest, and found a place in the same building, as it does in the present Guildhall. Wade was among the first donors to the fund for building the Mineral Water Hospital; he subscribed £100 when the Blue Coat School was erected; he gave handsome sums to the Corporation for public distribution; and 500 guineas towards a scheme for improving the Pump Room, but which eventually assisted in rebuilding St. Michael's Church on condition that a pew therein should be reserved for the Corporation. He was also a munificent donor to the Abbey, giving it a marble reredos, a stone font, and an altar-piece—"the Wise Men's Offering." The most important service rendered by him to the church was the building of Wade's passage, which, providing an access to the Grove, prevented the further use of the north aisle of the Abbey as a short cut to the same spot. His last charitable disbursement was in 1747, when he gave a liberal donation towards the building of a market for butchers. Wade died on the 13th March, 1748, aged 75, and was buried in Westminster Abbey, a fine monument, by Roubiliac, being erected there to his memory. Ralph Allen also raised a memorial to him in the grounds of Prior Park; it was an obelisk standing on a pedestal, on the sides of which were depicted in bas relief the General's more noted achievements. All trace of it disappeared many years ago.

The vacancy created by the death of Wade was supplied by the election of Sir John Ligonier, a more distinguished soldier even than his predecessor. Of Huguenot descent, he was born at Castres, in the South of France, in 1680, where he remained until the revocation of the Edict of Nantes, when, like so many of his co-religionists, he abandoned his native land rather than his creed. His brilliant military career furnishes another illustration of the great loss of talent France sustained by the arbitrary treatment of his Protestant subjects by Louis XIV., and how much England has gained in the departments of science, art, literature, and commerce, and in arms, through the asylum it afforded to the victims of his short-sighted policy. John Ligonier became commander-in-chief of the British Army, and two of his brothers, who followed him into exile, died, one a Colonel and the other a Major, in the service of this country. During his career, Sir John took part in twenty-two general actions and nineteen sieges without receiving a wound. His last engagement was at Val, where he led a brilliant cavalry charge of Scots Greys and Enniskillings, which saved the Duke of Cumberland and his retreating infantry from the French horse. Ligonier was among the prisoners taken; and Marshal Saxe, in presenting him to Louis XV., said, "Sire, I present to your Majesty a man who by one glorious action has disconcerted all my plans." The King would like to have secured the services of the hero, but he remained true to his adopted country, to which he returned when in his 67th year. In March, 1748, he was chosen to succeed Wade, though only by a majority of one, his opponent being Joseph Langton, of Newton Park, who received fourteen votes against fifteen cast for his rival. The choice was not between Whig or Tory, Hanoverian or Jacobite, but between a neighbour who was personally well-known, and a stranger who was known only by name and fame. The preference probably was given to Ligonier from a belief that he would prove another Marshal Wade, with whom the Corporate body had every reason to be satisfied. On his arrival in Bath, in January, 1749, he was met by Ralph Allen, who conducted him and his retinue to Prior Park. The next week Sir John, having been elected, entertained the Corporation, and, before leaving for London, he ordered £50 to be distributed among poor housekeepers, who had no relief from the parish. He was re-elected in 1754 and in 1761; but in 1766 he retired, being created an English Earl by letters patent, and having a seat in the House of Lords. He had previously received the honour of an Irish Viscountcy. Among other posts held by him was that of Lieutenant-General of Ordnance, from which he was removed by a political intrigue—an intrigue meanly countenanced by the Duke of Cumberland, then Commander-in-Chief, whom he had saved at Val. The Earl died on the 28th April, 1772, in his 92nd year.

At the General Election in 1747 Wade had (as stated) for his colleague Robert Henley, the scion of an old Somersetshire family, Langton being then also an unsuccessful candidate. Henley sat with Ligonier until 1757. He was a handsome man, of conspicuous mental ability, but with a mercurial disposition. With him the indulgence of his convivial propensities was much more congenial than the cultivation of his undoubted talents. His conversational powers, seasoned as they were with wit and anecdote, made him a pleasant boon companion, and a favourite in Bath society. He was born in 1708, and, having been entered of the Middle Temple, he was called to the Bar in 1732, and joined the Western Circuit. Before he had obtained renown as an advocate he was appointed Recorder of Bath, for which he was indebted to his social qualities. At this period the smart junior barristers were wont to pass their vacations in this city; and Henley was there the gayest of the gay. He distinguished himself among the ladies in the Pump Room in the morning, as well as among the topers at the Bear Inn in the evening. From his rattling, reckless disposition, matrimony seemed to have no attraction for him; yet, to the surprise of his friends, he formed a romantic attachment. To Bath came, for the benefit of the waters, a very young girl of exquisite beauty, who, from illness, had lost the use of her limbs so completely that she was only able to appear in public wheeled about in a chair. She was the daughter and coheiress of Sir John Huband, of Ipsley, in Warwickshire, who was the last male of a time-honoured race, whom Dugdale states to have been lords of the above manor in lineal succession from the Conquest. Henley, struck by the charms of her face, contrived to be introduced to her, when he was still more fascinated by her conversation. His admiration soon ripened into a warm and permanent attachment, which he had reason to hope was reciprocated. But it seemed (says Lord Campbell in his biographies of the Chancellors) as if he had fallen in love with a Peri, and that he must for ever be content with sighing and worshipping at her shrine. Yet, in course of time, the waters procured so effectual and complete a cure that Miss Huband was able to comply with the custom of the place by hanging up her crutches to the nymph of the spring and to dance the "Minuet de la Cour" at the Lower Rooms with her lover. Soon after, with the full consent of her family, she gave her hand to the suitor who had so sedulously attended her. To the end of a long life she continued to enjoy a perfect state of health, and their affection remaining unaltered, she gave him that first of human blessings—a serene and happy home. In November, 1756, Henley succeeded Murray as Attorney-General to the Crown, and in the following month, on the issue of a new writ, he was re-elected by the Corporation; but in the next year, being made Lord-Keeper, he vacated his seat, and in 1760 he was elevated to the dignity of Lord Chancellor and created Earl of Northington. He was a frequent visitor to Bath with his wife, and led a comparatively sober life, although his devotion to port remained without much abatement, and severe attacks of gout ensued. When hobbling to his seat in the House of Lords he was, on one occasion, overheard muttering to himself: "If I had known these legs were one day to carry a Chancellor, I'd have taken better care of them when I was a lad." He held the Great Seal nine years, and during four administrations. In his last illness, at his seat, "The Grange, Hants," he was reminded of the propriety of receiving the consolations of religion. He readily agreed that a divine should be sent for; but when the Right Rev. Dr. ——, with whom he had formerly been intimate, was proposed he said, " No, that won't do; for the greatest sin I have to answer for was making him a Bishop." A clergyman of the parish was substituted, and his ministrations were gratefully accepted by the dying man. His wife, by whom he had eight children—three sons and five daughters—survived him several years.

The successor to Henley in the representation of the city was the Great Commoner, William Pitt, elected in 1757. Prior to this Pitt had become a resident of Bath, having purchased No. 7 in the Circus, which was then in course of erection by the younger Wood. He was re-elected in March, 1761, his coadjutor being Lord Ligonier. The relations between Pitt and the Corporate body which elected him were of the most friendly character. He was given the freedom of the city in a gold box, and, in 1760, he and his colleague received a letter from it thanking them for the great services they had rendered the King and the country. Pitt, in his reply, assured the members that he was "justly proud of the title of servant of the city of Bath, and that I can never sufficiently manifest the deep sense I have of your distinguished and repeated favours, nor express the respect, gratitude, and affection," which he entertained for the Corporation. As a further proof of this mutual goodwill, his portrait, painted by Hoare, was placed in the Guildhall. Strange is it that a single word sufficed to sever these strong ties of reciprocal regard. So it was. At the close of the war with Spain, early in 1763, the peace concluded was bitterly denounced by Pitt; but it was approved by the Bath Corporation, who presented an address to the King, in which his Majesty was thanked for the "adequate" and advantageous peace obtained. Pitt took umbrage at the word "adequate"; the peace was, he said, repugnant to his fixed opinion of some of the conditions; he refused to accompany Sir John Sebright, his colleague, as requested, at the presentation of the loyal address, and intimated his intention not to seek re-election. The letter was addressed to Ralph Allen, who wrote a conciliatory answer, in which he took the entire responsibility for the wording of the address, including the offensive "adequate," and at the same time entreating Pitt to re-consider his decision, pending the receipt of which he withheld the missive from the civic authorities. But Pitt would not be pacified. He could not think of embarrassing and encumbering friends "who view with approbation measures of an administration founded on the subversion of that system which once procured me the countenance and favour of the city of Bath." Allen, finding the great statesman inflexible, in a dignified but feeling letter informed him that. "with the greatest anxiety and concern, I have, in obedience to your positive and repeated commands, executed the most painful commission that I ever received." Pitt gave practical proof of his resentment; he sold his house in the Cirous, and though he retained his seat until 1766, he only visited the city once in the interval, and that to take the waters. There was no need for him to seek another constituency, as he was created Earl of Chatham in the above year. In one of his letters to Horace Mann at this time, Horace Walpole writes, "My recovery has gone on fast; the Bath waters were serviceable to me; I left Lord Chatham at Bath in great health and spirits." Ralph Allen, in his will, gave striking proof of his unalterable esteem for his friend. A codicil says: "For the last instance of my friendly and grateful regard for one of the best of friends, as well as the most upright and ablest of Ministers that has adorned our country, I give to the right honourable William Pitt the sum of one thousand pounds to be disposed of by him to any of his children that he may be pleased to appoint."

The next Recorder after Henley was Charles Pratt, to whom, as Lord Camden, the reader has been already introduced. He was the third son of Sir John Pratt, and was born at Kensington in 1714. At Eton, where he was educated, he had as one of his contemporaries, William Pitt, afterwards the Earl of Chatham, with whom he was a life-long friend, and to whom he probably owed his official connection with the Bath Corporation, which commenced in 1759. As a barrister he rode the Western Circuit for some years without obtaining a brief, and, despairing of success, he thought of abandoning his profession and entering the Church. From taking this step he was deterred by the persuasion of friends, and subsequently chance or design threw the required opening in his way. He was briefed as junior to Robert Henley, who fell ill or feigned illness, and left him the entire conduct of the case, in which he displayed such conspicuous ability as to establish his reputation. He succeeded Henley in all his posts, including the Lord Chancellorship, and it was while in the enjoyment of these high honours that he figured so largely in Bath society. In the vexed constitutional points which the Wilkes' case engendered, and in other questionable proceedings by the Crown, Camden's decisions were generally given in favour of liberty. His popularity was great in consequence. The London Corporation presented him with the freedom of the city, and commissioned Sir Joshua Reynolds to paint his portrait. He received the freedom of other cities and the thanks of several Corporations, including that of Bath, while a full-length portrait of him, by Hudson, was hung in the Guildhall, Exeter; a third likeness was painted by Gainsborough; and another full-length, by William Hoare, hangs in the Bath Guildhall.

His son, the Hon. John Jefferys Pratt, who was born in 1759, and trained for the Bar, was elected one of the members for Bath in 1780, in conjunction with Abel Moysey. He continued to represent the city till 1794, when he succeeded to the Earldom of Camden. In 1812, he was created Earl of Brecknock and Marquis Camden, and from that date he was Recorder of Bath down to the passing of the Municipal Reform Act in 1835. As a reward for his father's services he was, in 1780, appointed one of the Tellers of the Exchequer, an office he held for the extraordinary period of 60 years—his death taking place in 1840. A futile attempt was made in 1812 to limit the emoluments accruing to him, which had increased from £2,500 per annum in 1782 to £23,000 in 1808. Camden, however, from that time declined to receive any income from this source, which at his death amounted to a quarter of a million sterling. He received formally the thanks of Parliament for his patriotic conduct. The political connection of the family with Bath was renewed in 1818, when the Earl of Brecknock, the only son of the Marquis, was chosen one of its representatives, the seat being retained by him till 1832.

Pratt's re-election in 1784 was remarkable from the fact that one of the candidates proposed was William Pitt, the son of Pratt's father's old friend, the Earl of Chatham. He was unsuccessful, although he had filled the office of Chancellor of the Exchequer in the Shelburne Ministry, and was now Prime Minister. It was generally thought that his hold of power was extremely precarious, and that he would soon be relegated to the Opposition benches. The Corporation sharing in this view, placed Pitt at the bottom of the poll, the numbers being Pratt, 27; Abel Moysey, 17; Pitt, 12. The unpopularity of the latter in Bath was largely due to the Shop tax. Pitt's financial reforms included a graduated tax on retail shops. Those rented from £4 to £10 were assessed at 1s. in the pound; from £10 to £15, 1s. 6d.; from £15 to £20, 1s. 9d.; from £20 to £25 and all above, 2s. in the pound. The "Bath Chronicle" of July 6, 1784, describes the indignation rife, and the peculiar mode, in some instances, of expressing it. The shopkeepers (it says) of this city, who will severely experience the effect of this greatly reprobated tax, did not shut up their doors yesterday, as most of them know they can ill spare a day, there being scarce enough in the year to keep their shops open on, to enable them to pay the accumulated burdens imposed on them. However, several shewed their contempt so far as to hang their windows with mourning crape, scarves, etc. Others covered their counters with velvet palls and hung weeping willows and other emblems of woe, with inscriptions expressive of their indignation against the tax and a once favourite Minister. "No Pitt," "No Partial Tax," and such like sentences were chalked up in every part of the city. A glazier put an old cucumber glass into mourning with a lamentation upon it for the loss of liberty and a declaration for ever against Pitt. Another shop had very strikingly reversed Magna Charta. A widow in Northgate-street exhibited, under a sable canopy, in her window the following lines:—

"Ill bodes the day, O Pitt! Severe thy laws.
When injured widows join their country's cause,
Mourn thy dire statutes, and consign thy name
To realms of darkness for thy field of fame."

Miss Pitcairn, at the tapioca shop in the Grove, erased for the day the first syllable in her name, and the following Hudibrastics appeared on her shutters :—

"The name of Pitt's so odious grown,
Tho't makes up one half my own,
Behold, I do renounce it truly
On this detested fifth of July;
And know from hence ('sans hoca poca')
That Agnes 'Cairn sells Tapioca."

With this feeling prevailing among the citizens it is not surprising that the municipal authorities chose the other candidates. They, however, made amends to the young statesman for his rejection : immediately afterwards he was presented with the freedom of the city in an elegant gold casket, with the customary laudatory address. Pitt was an occasional visitor to Bath, and his letters written hence show that he was much better for his sojourn here. He enjoyed rest and quiet and intercourse with Wilberforce, Canning, Rose, and other political friends, who were either regular or special visitors. During his last stay he received news of Napoleon's decisive victory at Austerlitz—a victory that shattered the great military combination which he fully expected would have put a stop to French conquests. Broken in health by his arduous duties and the anxieties inseparable from the crises which the French Revolution produced, his end, which speedily followed, was thought to have been hastened by the depressing effect of the news borne from the Austrian battlefield. The Wiltshires of Shockerwick preserved a curious reminiscence of this momentous event. Pitt, it is said, had gone there to see the Gainsborough pictures which then adorned the walls of the mansion. He was looking at one of them, when the sound of a horse galloping furiously up the gravel path leading to the house was heard. "That must be a courier," said the statesman, "with news for me"; and almost immediately a man, booted and spurred, and splashed from head to foot, entered the room and handed despatches to the Minister. Tearing them open he became intensely agitated and exclaimed, "Heavy news, indeed! Get me some brandy." Wiltshire's son rushed away and brought in the brandy, which was poured into a glass, and a little water added, when Pitt tossed it off. He took another tumblerful in the same eager manner, for had he not done so the opinion was that he would have fainted on the spot. Austerlitz was fought on the 2nd December, 1805, and Pitt died on the 23rd January, 1806.

Of the other candidates, successful or not, who passed before the Corporation during the latter part of the century no particular mention need be made, beyond the fact that, in 1790, Lord Viscount Weymouth was elected with the Hon. J. J. Pratt ; the former again in 1794 with Sir Richard Pepper Arden ; Lord Bayham and Lord John Thynne in 1796. From the foregoing memoirs, it will be admitted (we repeat) that the Corporation are justly entitled to commendation for the manner in which they exercised the trust confided in them of choosing representatives for Bath. As a rule, they took care to select the ablest and best men that could be found—men who conferred lustre on the constituency either from their talents or achievements, and benefited the State at large by their maintenance of the Hanoverian dynasty, or the support they gave to Constitutional principles when these were in danger. Hence it is that the city can boast of having numbered among its representatives two of the most distinguished generals of the age, the greatest of English orators and statesmen, two members of the Bar who attained the highest professional honours, and a third (J. J. Pratt) who was a member of three Administrations and for a time Lord Lieutenant of Ireland.

THE LOYAL ASSOCIATION.

In the last decade of the century the nation was confronted by great perils both at home and abroad. The writings of Tom Paine, William Godwin, and other doctrinaires found many sympathisers among political reformers; while the French Revolution, sanguinary within and aggressive without, threatened to engulph the whole of Europe, the defence of which devolved mainly upon this country, though the magnitude of the responsibility it had undertaken was not then fully realised. The cause of law and order on this side of the Channel was, however, strengthened by the terrible excesses of the Parisian mob. The horror these inspired made the more thoughtful dread any attempts at improving the British Constitution, the maintenance of it in its integrity being deemed a patriotic duty. The feeling thus engendered caused an outbreak of loyalty and devotion to the Throne similar to that witnessed throughout Great Britain and her Colonies in defence of the interests of the Empire, when they were jeopardised by the late Boer War. The prevailing opinion was expressed through associations formed for "preserving liberty, property, and the Constitution against republicans and levellers." Such an association was established in Bath. Among the archives of the Corporation is a large folio volume giving the particulars of the organisation. It is dated "Guildhall, December 8th, 1792," and the declaration of allegiance is prefaced by a statement that "divers inflammatory and seditious publications have been industriously dispersed, and clubs and societies instituted, by factious and evil minded men." This is followed by a confession of the "peculiar happiness which this country enjoys of a free Constitution, so well and so wisely framed as to have stood for ages the envy and admiration of Europe," and of alarm "at the violent convulsions, anarchy, and massacres of a neighbouring Kingdom occasioned by the attempt to reduce into practice the delusive and visionary principles recommended in the seditious publications above alluded to." Hence it had become the interest as well as the bounden duty of good and loyal citizens of all ranks and degrees "to unite and publicly avow their attachment to the King and our excellent Constitution." Three resolutions embody the sentiments above sketched, while a fourth explains that the object of the Association is "to give support to the civil magistrates in all emergencies that may arise."

About one thousand signatures are appended, not only the names, but the addresses and occupations (if any) being given. Pasted in is a list of eighty-one residents of Stoneaston who, sympathising with its objects, wished to join the Bath Association. The chairmen of the city also avow their loyalty in a separate memorial addressed to the Mayor (Mr. Abel Moysey). "We think it our duty (they say) to make a tender of our services in this voluntary manner. We are conscious that our livelihood and the happiness of ourselves and families depend entirely upon the prosperity and peace of the Kingdom in general and of this city in particular. We therefore thus enrol our names and state our residences in order (which God Almighty may be pleased to avert!) if necessity should ever require it, that we may be called upon to give our utmost support to your laudable undertaking. We love and honour our King and are ready to lay down our lives to serve him, to protect his laws, and to obey the commands of all who are in authority under him." There are 164 names attached, all in pairs, showing it was the day of Sedan chairs. The deferential tone of the above address may cause a smile of contempt in the present day when respect for superiors, and reverence for things sacred and secular, have become almost obsolete virtues among masses of the people; but it by no means follows that real manliness is fostered, that the social fabric is strengthened, or that the industrial supremacy of the country is secured by their disappearance, especially when the conceit of self-assertiveness take their place. According to the official arrangement the book remained open at the Guildhall for one week, three members of the Committee attending from 12 to 3 to receive names. If the limit prescribed was not exceeded it shows how enthusiastic must have been the loyalty of the city and its suburbs that such a long array of signatures should have been obtained in so short a time. It does not appear that the Association did anything beyond enrolling members, the necessity for carrying its principles into practice not arising. At the same time, its formation helped to give an impetus to the Volunteer movement, which was called into existence a few years later by the hostile designs of Napoleon against this country.

Lacock Abbey.

CHAPTER VII.

THE HIGHWAYS AND THE POST OFFICE.

S the aim of this work is to give a complete survey of the progressive movements in the last century, it may be necessary to traverse ground familiar to some of its readers. The services of Ralph Allen and John Palmer come under this head, as everyone has heard of the reforms effected by the two distinguished citizens—reforms which made the sending of letters certain and expeditious, and travelling more easy and safe. The mail coaches introduced by Palmer not only secured speedy intercommunication, but likewise helped to clear the roads of the highwaymen by whom they were infested, and to accelerate the passing of Turnpike Acts for their improvement. In this railway age people have little conception of the dangers which faced travellers, both from hills that had to be descended, steeps that had to be negotiated, and the thieves lurking in lonely haunts ready to pounce upon them. We have only to look at Holloway, as it is now, under improved conditions, to see the difficulties that vehicular traffic had to contend against. Think what must have been the anxieties and trials of passengers when dragged up this abrupt and narrow thoroughfare, and what the cruelty to horses attached to heavily-laden coaches! Bad as it was here, it was still worse in the descent to or the ascent from Midford. Anyone can judge of this who, instead of following the road skirting the Midford Castle grounds, takes the one a little to the right. He will find himself in a lane narrower and more precipitous than Holloway, extending to the hamlet, and bi-sected midway by the modern road. No wonder that old Lelande notes at this point, in his quaint way, that it was all "by mountaine and quarre" to Bath; he must indeed have fancied himself climbing a mountain side, as, through long years after, did the drivers of the stage-wagons, "the flying machines," and the private coaches of the nobility and gentry and their occupants, when taking the same route. Other roads on the Bathford side, and elsewhere, were equally like escarpments. Nor was this the only cause of complaint. These were the days before McAdam had introduced his improved method of making and mending the King's highway. The roads, such as they were, were repaired in the rudest possible fashion, adding, by the joltings and stoppages occasioned, in consequence of ruts and stones, greatly to the misery and weariness of travellers. Their condition, moreover, was not due to lack of statutory powers. Enabling Acts were passed by Parliament; but the works authorised were wholly neglected, or done only partially. The need of making the highways better was brought home to the public mind during the visit of Queen Anne, whose carriage narrowly escaped accident when trying to climb the heights of Lansdown. The £1,800 raised and expended in the improvement of the road afterwards was doubtless due to this alarming incident. It seems also to have convinced the authorities that the other thoroughfares leading to and from the city were not as safe as they should be, and a Bill was, without delay, drafted for their reparation, and other matters. It was entitled "an Act for repairing, mending, and enlarging the highway between the top of Kingsdown Hill and the city of Bath; also several other highways leading to and through the said city, and for cleansing and lightening the streets, and regulating the chairmen thereof." The preamble is laudatory: "Whereas (it runs) the city of Bath is a place of very great resort from all parts of the Kingdom of Great Britain, and from foreign parts, for the use and benefit of the waters and drinking the mineral waters there;" it then schedules the principal roads, beginning with that leading to Kingsdown, and, to the intent that the same may be forthwith effectually repaired and hereafter kept in good repair, it ordains that Commissioners shall be appointed, the same to be two Justices of the Peace of the nearest parts of the counties of Wilts, Somerset, and Gloucester, and one from the city of Bath, to carry out the Act. Surveyors residing near the respective ruinous places, after being chosen, were to report to the Commissioners, who were empowered to order work to be done, to raise tolls for the payment, and to borrow money to the extent of £3,000. The Act was passed in April, 1707, and was to remain in force for 21 years. Nothing was done under its provisions beyond a little cobbling here and there; and at the end of seven years another Act had to be obtained, because the highways were not properly repaired, and because the Commissioners were in debt £2,400, borrowed money, and not being able to pay either the interest or the principal, the creditors, it was confessed, had taken possession of the tolls, and left them penniless. Parliament again came to the rescue—extended their borrowing powers, and authorised them to assign the tolls as a security. Once more money was obtained from both these sources, but so far as the roads were concerned there was no appreciable improvement in their condition. The hill, for instance, at Kingsdown, which occupied the chief position in the preamble, was in its upper part so steep, and so rough, that strangers coming to Bath were filled with fear and terror "while vagrants were guiding their coaches, as though they were in the most imminent danger of overturning and tumbling down a precipice." Dissatisfaction was universal; but the muddling went on till 1754, when, there being a debt of £3,000, and no funds as usual, a third Bill was promoted to augment the tolls and duties, and to embrace other roads in the district previously omitted. Parliament seems to have had enough of the inentitude of the parochial mind, and, as a corrective, created a body to carry out the new Act large enough for a County Council. It appointed as Highway Commissioners about one hundred of the principal landowners of Somerset, Gloucestershire, and Wiltshire, whose names are set out in detail in the Statute—all the members of Parliament for the three counties, the Rector of Bath, the Mayor, Recorder, Aldermen, and Common Council for the time being of the city. However much shirkings may neutralise the power of numbers, or evasions result from too wide a diffusion of responsibility, the plan adopted in this case was decidedly successful. Though amending Acts were subsequently found to be necessary, from this time began the thorough re-modelling of the highways. Gradually the rough was made smooth, the crooked straight, the narrow widened, and the precipitous reduced or circumvented.

Turning now to the "Knights of the Road," the following exploits, culled from the records of the time, will illustrate their activity and mode of operation:—In March, 1753, Dr. Hancook, of Salisbury, was coming over Claverton Down with his daughter, eight years of age, in a postchaise, when they were attacked by two highwaymen, who fired into the carriage three or four times; one bullet passed over the child's head, the others shivered the glass to pieces. The chaise then halted, and the villains took from the doctor thirty-six shillings and his gold watch, and carried off all the luggage. They threatened also to kill the child to make the father disclose if he had bills or any more valuables about him. The Bath machine, when returning to Salisbury on the afternoon of September 17, 1759, was stopped near the ten-mile stone by a single highwayman, disguised with crape over his face, who robbed every one of the passengers, four in number, of their money and rings, and the coachman of fifteen shillings. " He behaved very bold and resolute, was rude in his manner, and detained the company half-an-hour." In May 1765, Mr. and Mrs. Palmer, of Drury Lane Theatre, and Mrs. Pritchard, the celebrated actress, left London in the Bath post-coach to fulfil an engagement here. They were stopped by a mounted highwayman, who, with pistol in hand, compelled Mr. Palmer to surrender forty guineas and a gold watch. Ten years later, in the last days of June, there was a small cluster of these crimes recorded. Mr. Symons, attorney of Bristol, while returning from Bath, was stopped by two highwaymen, with loaded pistols, near Keynsham, and robbed of half-a-guinea and some silver. A flying machine from London to this city was robbed on Hounslow Heath by a single highwayman, and all the passengers had to pay heavily before they could get rid of the scoundrel. A flying diligence was stopped near Batheaston by a single highwayman, indifferently mounted, who made a collection from the occupants; and the next night he rifled two postchaises near the same place. The Rev. H. Harington, uncle of Dr. Harington, of Bath, and the Rev. Mr. Birch, going from Devizes to Salisbury, were intercepted by a highwayman on the Plain. He presented a pistol at them, and demanded their money. Mr. Harington took three guineas from his purse, but the fellow said his necessities were great and he must have more. Mr. Harington gave him an additional couple of guineas, on which he asked for Mr. Birch's purse; but on being told that it contained a remarkable ring that might lead to the discovery of the robbery, he declined to take it and rode off. A post-chaise and four, in which were three ladies and three gentlemen, was brought to a standstill at the bottom of Saltford Hill by a single highwayman, who presented a large pistol at one of the gentlemen, and resolutely demanded money. He was requested to withdraw the pistol and not terrify the ladies, with which he complied, and then collected about thirty shillings from his victims. On leaving he shook hands with the same gentleman, and wished them all " Good night," at the same time telling them that in case of another attack they were to say that " Young Turpin" had been with them! As no mention is made of a similar visitation it may be concluded that the travellers finished their journey without further molestation. So great was the terror rife from these outrages that the farmers attending the Bath Market were wont to start for home in groups for mutual protection. The same activity was displayed all over the country by footpads, who roamed the main roads and plundered carriage people and others with impunity, there being no constabulary to check their raids. The gibbet, to which some of them, when caught, were doomed had no terrors for them; their sole ambition was "to die game." Even women took to the road, and not a few ruined gamesters, who, in this way, hoped to mend their broken fortunes.

These facts will enable the reader to judge of the conditions governing the conveyance of passengers and goods from place to place, and of the difficulties confronting those who initiated improvements. If, as a rule, few people left their native place, or if before undertaking a journey their wills were made, such stationary habits will not be regarded with wonder, nor the prudent settlement of their affairs be deemed superfluous. The pioneer of beneficial change in the transmission of letters was Ralph Allen. He was not a talker, but a doer—taciturn and thoughtful, a close observer of men and things, and, withal, possessed of a fine talent for organisation. His experience in the subordinate work of the Post Office, as well as his position of postmaster of Bath, had made him familiar with the chaos and peculation prevailing throughout this department of the public service, and the utter failure of the attempts made by the Government to secure order and honesty therein. In those days, as now, Royal Commissions were appointed to make inquiries and suggest remedies for the correction of the abuses. The reports of these bodies could not but condemn evils so apparent everywhere. Corruption reigned supreme; the public were systematically cheated; the treasury was as systematically robbed to enable the officials to enrich themselves. As early as 1710 (the year in which Allen was transferred to the Bath Post Office) an Act was passed for the establishment of a General Post Office throughout her Majesty's dominions. Its preamble recites "that divers deputy (or country) postmasters did then collect great quantities of letters, and by clandestine and private agreement among themselves did convey the same in their respective bags without accounting for the same, to the great detriment of her Majesty's revenue." Heavy penalties were enacted; but the iniquitous practices went on unchecked. Six surveyors were afterwards appointed to investigate, and devise improved methods to safeguard the public interest. The reports of these roving officials were voluminous enough, but not the shadow of a practical suggestion could be found in any of them—at least nothing on which the Postmaster-General felt himself justified in acting.

Yet anything more primitive and faulty than the way letters were dealt with could hardly be conceived. All letters that did not go to or pass through London were called bye or way and cross-post letters, and these were thrown promiscuously together into one large bag, which was opened at every stage by the deputy or any assistant, who threw the whole of the contents into a heap, picked out what belonged to his delivery, tossed the rest into the bag, and sent it on its way, the same process being repeated until all the letters were exhausted. Is it surprising that under such a rough-and-tumble treatment many missives were lost? Nor was this the only cause of miscarriage. As the postage was prepaid, it was a common practice to destroy letters and appropriate the postage. The scandals rampant are illustrated by the fact that one of the surveyors above mentioned records it as most singular that Mr. Allen, of the Bath Post Office, had scrupulously accounted to the Government for the postage received from the day he entered on his appointment. In addition to the frauds thus committed, there were no postmen regularly engaged to deliver the letters. Loafers hung about the Post Office door, like touts are wont to do at railway stations, and to them were entrusted the finding out and the delivery of the letters to the persons to whom they were addressed, for which service a gratuity was expected.

Allen undertook on his own responsibility to introduce a fresh scheme—one that would ensure a more rapid conveyance of correspondence, and add to the revenue. He was asked by the Postmaster-General to disclose his plan, which he declined to do; but he offered to defray all expenses attending its working and to guarantee a gain of £2,000 to the Government. All kinds of objections were urged against him, and it was not until he had promised to make good any deficit in the "country letters"—a branch the Government had retained in its own hands—that a contract was concluded (1720) for seven years, leaving him to manage the "by way, and cross post" letters. Allen's plan was the employment of mounted post boys to carry letters, and the organisation of a trustworthy staff to receive, dispatch, and deliver them. The all-round improvement effected by his system was so apparent that he found no difficulty in getting, at the end of the first seven years, a renewal of his contract for a like term. Altogether it was renewed six times, but not without a struggle and fresh concessions from the contractor each time. His net profit in 1761, as shown by a return he made to the Postmaster-General, was £12,248 7s. 1¾d. per annum, while the aggregate gain to the Treasury from 1720 was £1,516,850. It has been said that Allen embarked in the stone trade to conceal the large profits he was making by his postal arrangements. The assertion is disproved by facts. He did not commence operations under his scheme, as we have seen, until 1720; and long before his first term was expired his tramway to Combe Down was complete, and the wagons were bringing stone to his wharf in the Dolemeads. His gains at this early period could hardly have been so large as to require concealment, nor could he have anticipated the splendid income that he eventually derived from his patriotic enterprise. It is remarkable testimony to his keen commercial instincts that he was able to conceive and carry out successfully two such large undertakings concurrently. Where he obtained the necessary capital from is a matter of conjecture. In his first application to the Government he described himself as a young man just entering into the world, and anxious to push his fortunes for his own and the public benefit. He had, however, married Miss Earl (as already stated), a natural daughter of Marshal Wade, and there is good ground for assuming that he was at the outset financed by his wealthy father-in-law, who appreciated his capabilities and probity. The citizens at large equally admired his character, in which benevolence was a conspicuous trait; and the Town Council, of which he was a member, deferred in all weighty matters to his judgment. In consequence it was caricatured as the "One-headed Corporation," to the great mortification of Allen, who is delineated as a giant among pigmies. He was content with the power he indirectly wielded, and the homage he received, without making them a stepping stone to the Parliamentary honours that he could easily have reached. His manners were unobtrusive, like those of a plain man, which was his customary pose. Yet he worthily played the part of a Mæcenas in his stately home to men of learning, and of a courtier to the princes and grandees who came under his roof. From some preceding incidental references it may, perhaps, be justly inferred that he had the common weakness for a lord; it may be equally true that he was not altogether free from "the pride that apes humility." But these were as mere spots on the sun's disc. His career as a whole was that of a high-minded, conscientious man, who accumulated wealth by rare shrewdness and tact, who benefited the nation by his administrative reforms and the city of his adoption by his enterprise and zeal for her interest, one of whom Pope could say that he "did good by stealth and blush'd to find it fame." Allen died in 1764 at Prior Park, the subsequent history of which tells the old story of the vanity of human wishes. The beautiful domain, the glory of its founder, who hoped to see it maintained, and the pride of his contemporaries, has, in the years since intervening, waned under neglect and decay, save during the brief period (1834-44) when Bishop Baines was its nominal owner.

A worthy successor to Allen as a reformer in the same direction was John Palmer, who possessed the like practical turn of mind and energy in grappling with evils which distinguished his predecessor. He probably inherited these qualities from his father, who, in addition to conducting locally the businesses of a brewer and tallow-chandler, dabbled in theatrical undertakings, in which his son assisted. On the retirement of the father in 1766 young Palmer became the sole lessee of the theatre. A great obstacle in his path was the slowness with which the "stars" he engaged travelled to fulfil their engagements and the risk to purse and life they had to face. He turned is attention to the removal of these difficulties, being doubtless stimulated by the knowledge he obtained of Allen's success financially. It was then that he formulated the scheme to substitute mail coaches for post horses; the coaches to convey both letters and passengers, and to travel eight miles an hour. At this time the "flying machines," as they were called, took from two to three days to accomplish the journey to London—"if God permits," which was significantly added to the advertisements, in view of impediments from bad roads and footpads. Each of Palmer's coaches was to have a guard perched on the outside armed with a blunderbuss to deal summarily with any enemies on the prowl. In October, 1782, the proposal was brought under the notice of Pitt, who referred it to the Post Office authorities, by whom, of course, it was pronounced impracticable. Palmer was not to be foiled; he worked and agitated, and, through the influence of Lord Camden, Pitt, two years later, arranged a conference between the promoter and the Post Office officials, who had reported adversely. As a result, the plan was ordered to be tried on the London and Bristol road, and on the 2nd August, 1784, Palmer superintended the departure of the first mail coach. The experiment proved a decided success; time was saved, safety secured. It had, however, still to encounter opposition. The district postmasters purposely delayed letters to cast discredit on the system; the innkeepers petitioned the Government against it, urging the injury they should sustain, and the probable loss the revenue would have to face. Complaints also were made that the discharge of firearms by the guards alarmed the dwellers near towns, and disturbed the tranquillity of villages. To put a stop to the latter grievance "the guards were instructed not to use their arms wantonly, but to be particularly vigilant in case of an attack."

In spite of this grumbling and fault-finding the system was rapidly extended, its obvious usefulness securing it many friends. The Bath Corporation strenuously supported Palmer, and sent the Prime Minister a memorial, thanking him for the countenance he had given to the project, adding, "We cannot but avow pride as well as satisfaction in the reflection that the only substantial reforms in a department of such great importance to this commercial country have been contrived and executed—the first by a member of this Corporation (Ralph Allen), and the present by a native of this city, as well as one of our Council." Bristol and other towns adopted similar memorials, for, by the Autumn of 1785 mail coaches were running not only to London, but to Nottingham, Birmingham, Worcester, Norwich, Liverpool, Manchester, Leeds, Exeter, Gloucester, Hereford, Swansea, Milford Haven, Shrewsbury, Holyhead, Portsmouth, Dover, and, in the following year, to Edinburgh. Palmer's cause was won, and he naturally expected to reap the fruit of success. He had been verbally promised, through Pitt's secretary, Dr. Prettyman, $2\frac{1}{2}$ per cent. on the increase in the Post Office revenue during his life, with the general control of the office and its expenditure; but delays arose in reducing the terms to writing, and Palmer had to content himself for the time with the post of Controller-General and a salary of £1,500 a year. Friction, however, soon arose between him and the Postmaster-General, owing to his masterful spirit. He opened an office for newspapers, appointed the staff, and fixed the salaries, without consulting his chief, to whose request for an explanation he vouchsafed no reply. Even when it was pointed out to him by Pitt that he had power to suspend, but not to appoint, Post Office servants, he paid no attention to the reminder. A friend, Charles Bonner, whom he had made Deputy-Controller, publicly accused him of supplying information to the chairman of a meeting called in London to protest against some delay in the delivery of letters consequent on the action of the Postmaster-General. Palmer suspended Bonner, and, declining to justify the step to his chief, was in turn suspended. Pitt was anxious that there should be a court of inquiry into the whole controversy. To frustrate this Bonner basely gave to the Postmaster-General (Lord Walsingham) some private letters which had passed between him and Palmer during their intimacy containing compromising allusions. In one letter, referring to Lord Walsingham, Palmer says: "Let him be bullied, perplexed, and frightened, and made apprehensive that his foolish interference may even occasion a rising of the mail prices at £50,000 per annum difference to the Post Office." When his Lordship talked of appealing to the Treasury, he writes to Bonner: "It is the very thing I ought to wish, and must end well; yet it revives old quarrels and fevers me, in spite of myself. D—— them! I never can be absent to get a little bathing or quiet, but this is the case." In a third he asks: "Did Bartlett mention to you that they had been telling their story to the King? Pretty masters! So they complain to Dominie of the great boy." On a fresh list of the establishment being issued Palmer found himself excluded, and Bonner occupying his place. At the end of two years this arch-foe was dismissed.

Though slighted by the Government, Palmer received abundant evidence of the gratitude of the nation. He was presented with the freedom of London, Liverpool, York, Hull, Chester, Manchester, Ennis, Aberdeen, Inverness, Gloucester, and other towns. Tokens were struck in his honour, and from the Glasgow Chamber of Commerce he received a large silver "loving cup," which, in 1875, his granddaughter presented to the Bath Corporation. With the country practically at his back, Palmer pressed for a recognition of his claims by the Crown. In 1793 he obtained a pension of £3,000 a year, which was very inadequate compared with what 2½ per cent. on the increased revenue would have yielded. Both he and his son (Col. Charles Palmer) worked long and hard to obtain more liberal compensation, which two committees of the House of Commons had recommended. At length, in 1813, Lord Liverpool's Government introduced a Bill for the payment to Palmer of £54,702 from the Consolidated Fund, without fee or deduction and without affecting his pension. The Bill passed both Houses, and thus ended a struggle, the maintaining of which cost the Palmers £13,000. Some time before this Palmer had sold his interest in the Bath and Bristol theatres, and also in a spermacetti manufactory he was running, to devote his energies exclusively to the postal reforms he undertook. He was elected a member of the Corporation in 1775, chosen Mayor in 1796 and 1809, and was returned to represent the city in Parliament in 1801, 1802, 1806, and 1807. Having accepted the Chiltern Hundreds in 1808, his son, the Colonel, was elected in his place. He died at Brighton on August 16, 1818, and his remains were interred in Bath Abbey in the presence of the Mayor and Corporation; a mural tablet marks the spot. Palmer's difficulties with the Government were mainly due to his own faults. His zeal was not tempered by discretion. Self-willed and obstinate, he chafed under restraint, particularly when it came from official superiors whose knowledge of his schemes was much inferior to his own and whose judgment on disputed points was perhaps as weak as his own was sound. Too haughty to conciliate, his natural irritability was increased when he found that he had made enemies instead of friends, as well as jeopardised rights, which, as shown, cost him long years of worry and a large outlay to get conceded. Unfortunately for him he had not the tact and urbanity of his predecessor. Allen for more than forty years contrived to work amicably with the heads of the department. Palmer in less than seven years was at enmity with them; found himself shelved from the public service and deprived of the emoluments he had expected. In the end the recompense he obtained from the Crown was not disproportionate to his services, though it fell much short of that which Allen received, and was less than his early anticipations evolved,

CHAPTER VIII.

LITERATURE AND LITERATI.

REVIEW of the literature of the era brings into prominence the works published by the Faculty on the Bath Waters. So many are they as to give rise to a suspicion that a medical man did not consider himself duly accredited to practice his profession here until he had written a book or treatise on the thermal springs. At least some thirty works attest the anxiety of the doctors to proclaim in print their intimate knowledge of the therapeutic uses of these waters, and thus to inspire the confidence of invalids. To enumerate the productions or to attempt an analysis of their contents would be tedious and unprofitable, though crude theories might be revealed with their amusing side. Doubtless a few of them were written apart from any selfish consideration, and with the sole object of scientifically explaining the phenomena of the local thermæ, or of more accurately defining the diseases that they were likely to benefit, and how best to employ them as remedies externally and internally. Where these conclusions were the result of actual and long experience, as well as close observation—as in the case of the two Olivers, W. Falconer, Lucas, Cheyne, Charlton, etc.—the empirical notions of older practitioners were superseded and more reliable knowledge obtained for general use.

Of the works of historic interest compiled here, partly or entirely, three stand out conspicuously. They are: Wood's " Essay Towards a Description of Bath," Collinson's " History of Somerset," and Warner's " History of Bath." The first of these, as the reader has already learned, is disfigured by the whimsical antiquarian theories of its author, but gratitude is due to him for much of real historic interest that he has preserved, as well as for the description he gives of the city and its suburbs during his professional career, and the government and laws by which its affairs were controlled and managed. The second work (three large quarto volumes) was the outcome of ten years of constant labour by the compiler and his coadjutor, Edmund Rack. The forty hundreds into which the county is divided, and the 482 parishes contained therein, are historically and topographically treated, and though later researches have disclosed some errors, the general accuracy merits hearty commendation. Warner was a ready and voluminous writer, but his fame rests chiefly on his " History of Bath," which cost him two years of research, showing his determination to be as far as possible trustworthy in detail. The work is methodically arranged, and its diction is clear and vigorous. It is, and is likely to remain, the standard history of the city. Another work, though of a very different class, deserves special mention, viz., " The New Bath Guide," by Christopher Anstey, who spent the greater part of his life in Bath, and gave to the world one of the most vivacious books ever written in rhyme, containing withal a vivid picture of Bath society as seen by a candid friend from a humorous point of view. Its easy, lively style and persiflage introduced a new kind of composition, which has since found many imitators, notably by Barham in "The Ingoldsby Legends."

The other literary celebrities connected with Bath either as residents or visitors are so numerous that we can do no more than mention the more prominent. These include the pious Robert Nelson, author of "Fasts and Festivals of the Church of England," who was one of the founders of the Blue Coat School in 1711; Daniel Defoe, who, in the year just given, was collecting local particulars for his "Tour in Great Britain," and who here met Alexander Selkirk, from whom he gathered the germs of his immortal work, "Robinson Crusoe." Sir Richard Steele whiled away much time in Bath, but not unprofitably, as his papers in the "Spectator," "Guardian," &c., attest. Among the health-seeking visitors in 1727 was Dr. Croft. the celebrated musical composer. He died here in August of that year, and was buried in Westminster Abbey. The poet Gay spent a portion of 1728 in Bath with the Duchess of Marlborough, and helped to nurse Congreve, the dramatist, who accompanied her Grace, in his gout. Pope came hither before Prior Park was finished, and more often after, when he had Allen's beautiful grounds to walk and muse in, as well as an arbour where he could court solitude; here he might have met or seen one of his pet aversions, Lady Mary Wortley Montague, who was a visitor, and in her playful lyric, "Farewell to Bath," writes:—

> To all you ladies now at Bath,
>   And eke, ye beaux, to you,
> With aching heart and wat'ry eyes,
>   I bid my last adieu.

Lord Chesterfield wrote hence some of those "Letters to his Son" by which his Lordship is chiefly remembered. Fielding hovered in and around Bath, studying life and embalming the fruit of his observation in "Tom Jones." His sister, Sarah Fielding, made her reputation as a novelist while living here; Fanny Burney (Madame d'Arblay) after being an occasional visitor became a resident, and some of her published letters were written in Bath. Smollett hoped to find professional success, but instead found materials for "Humphrey Clinker." Harriet and Sophie Lee, while residents, wrote their "Canterbury Tales" and other works. Mrs. Radcliffe was attracted hither to recruit in the interval of publishing her thrilling and romantic stories. Miss Austen reaped the harvest of a quiet eye in Bath, where she was busy with her pen. Sheridan started as an author here, and sketched, from the motley crowd he met, the characters delineated in his sparkling comedies. Shenstone of the Leasowes was not averse to the gaieties of Bath society. Bishop Warburton, who owed his "lawn sleeves" more to the influence of Allen than to the popularity of his own "Divine Legation of Moses," dwelt at Prior Park after the death of Allen, one of whose nieces he married; Johnson and Boswell were among the pilgrims to Bladud's shrine in 1778; Goldsmith was a frequent votary, either as a valetudinarian or as the guest of Lord Clare, and gathered on the spot facts for his "Life of Beau Nash"; Cowper, much as he loved seclusion, was drawn by the fame of Bath or his interest in Lady Hesketh, his cousin, and left a record of his visit in "Verses on finding the heel of a shoe at Bath"; Edmund Burke came for health, and found a wife, the daughter of Dr. C. Nugent; Richard Graves, Rector of Claverton, author of "The Spiritual Quixote," was long a noted figure in local literary circles; Dr. W. Falconer, an eminent resident physician, is said to have published no less than 45 treatises; William Melmoth, translator of "Pliny's Letters" and some of the works of Cicero, was a citizen; as was also Malthus, author of the "Essay on Population." A few of Horace Walpole's gossiping epistles are dated from Bath, his place of occasional residence; De Quincey's early education at the Grammar School entitles him to a place among local celebrities; Southey passed his childhood here, and records his juvenile yearning to get to Sham Castle, which was a feature in the landscape as seen from his home in Walcot Street; Coleridge preached in the Unitarian Chapel, Trim Street; Mrs. Thrale, as Mrs. Piozzi (her marriage to Mr. Piozzi took place at St. James's Church), was long a resident and so for a shorter period was Hannah More; Sir Philip Francis, the suspected author of Junius's letters, was a frequent visitor to his father, Dr. Francis, who was an inhabitant; Gibbon, the historian, spent some time with his step-mother, who lived in Belvedere, and thought at one period of making Bath his permanent residence; Wilberforce again and again sought repose in Bath; Wilkes came to see his sister, Mrs. Jefferys, whose social peculiarities, as described by Warner, were as harmless as her brother's were not; William Godwin, the author of "Caleb Williams," was the wooer of Harriet Lee shortly after the death of his first wife, but dislike of his socialistic and religious principles made the lady obdurate, though she corresponded with him after he had left Bath. Dr. Maclaine, translator of "Mosheim's Ecclesiastical History," passed his closing years here, and was buried in the Abbey; Governor Pownall merits remembrance from the interest he took in the first discovery of the Roman Baths when the Priory was pulled down, and the record he left of the same. He died in Bath in 1805. Dr. Percy, Bishop of Dromore, and the author of the "Percy Reliques," was an intimate friend of Dr. Harington, whose acquaintance he made locally; Mrs. Delaney knew Bath well, as her memoirs disclose; Mr. Skrine, of Warleigh, published his "Rivers of England" in 1783; also "Tours in Wales" and "The North of England." Beckford, of "Vathek" fame, first met at the Assembly Rooms Lady Margaret Gordon, whom he subsequently married, and to whose early death he paid the homage of a life-long widowhood. William Hone, of the "Every Day Book," was born in Bath in 1780. The 18th century saw the birth of a local press. The "Bath Journal" was established in 1742, the "Bath Chronicle" in 1757, and the "Bath Herald" in 1795.

CATHERINE MACAULAY.

The name of this lady is omitted from the foregoing list because of sundry singular circumstances connected with her residence here, which deserve description. She was the widow of Dr. George Macaulay, prior to whose death in 1766 she had commenced publishing a "History of England" from the accession of the Stuarts, the advanced opinions expressed in which exposed the work to vehement attacks and the authoress to be denounced as a Republican. She came to Bath in 1774, and took a house on St. James's Parade, where she made the acquaintance of the Rev. Thos. Wilson, D.D. (son of the Bishop of St. Asaph) and the non-resident rector of St. Stephen's, Walbrook, London. At his request she removed to his residence, Alfred House (No. 2), Alfred Street. In return for this hospitality she began "A History of England from the Revolution to the Present Time," in a series of letters to Dr. Wilson. Only one volume was published. In it she favours James II. more than William III., while she denounces a standing army as "that unconstitutional engine of despotism," and the creation of a National Debt as "a diabolical engine which has long threatened to put an end to the prosperity of our country." Whether from liking her principles or her person, her host's admiration for her increased extravagantly and manifested itself foolishly. Indeed his actions deflected so far from the rational as to justify a suspicion that he was not free from cerebral weakness. Thus in the spring of 1777 "he erected in a marble niche or recess, properly decorated, within the chancel of St. Stephen's Church, a superb white marble statue representing Mrs. Macaulay in the character of History, in a pleasing antique style." She held a pen in one hand, and in her left a scroll bearing a quotation from her own "History"—" Government is a power delegated for the happiness of mankind when conducted by wisdom, justice, and mercy." Another inscription spoke of her as "a kind of prodigy," one who should " taste the exalted pleasure of universal applause." Space was left at the top for an epitaph, and under it was the following record,—" Erected by Thomas Wilson, D.D., rector of this parish, as a testimony of the high esteem he bears to the distinguished merit of his friend, Catherine Macaulay. A.D. 1777." A vault was made near to receive the remains of the lady in due course.

While the statue was in the hands of the artist, Alfred House was the scene of a novel demonstration. The 2nd of April was Catherine's birthday; it was ushered in by the ringing of bells, and was brought to a close by an " elegant entertainment " given by Dr. Wilson. Mrs. Macaulay, very elegantly dressed, was seated in a conspicuous, elevated situation in front of a numerous and brilliant company, when six poems and odes, which had been composed for and presented on the occasion, were read, with great propriety and expression, by six gentlemen selected out of the guests. One ode, it is said, was delivered with a grace and elocution that would have done honour to Garrick, and another with an energy and action not unworthy of Demosthenes. At the close of these poetic offerings, " that honour to the Church and human nature, the pious, learned, and patriotic Dr. Wilson, advanced and presented Mrs. Macaulay with a large and curious gold medal, struck in the reign of Queen Anne and given by her Majesty to one of the plenipotentiaries at the peace of Utrecht; which he accompanied with a speech expressive of her merit and of his friendship and veneration. Next advanced the ingenious Dr. Graham, to whom the world is so much indebted for restoring health to the guardian of our liberties and thereby enabling her to proceed with her inimitable history. He with great modesty and diffidence presented her with a copy of his works containing his surprising discoveries and cures," to which he prefixed a dedication reeking of the fulsome and vain-glorious. Mud baths were his speciality; and at one time he travelled with the beautiful Emma Lyons, afterwards Lady Hamilton and Nelson's " Emma," whom he exhibited reclining on a curious " celestial bed," as the goddess of health and beauty. Graham claimed " the supreme blessedness of removing, under God, the complicated and obstinate maladies your (Mrs. Macaulay's) fair and very delicate frame was afflicted with." At the conclusion of these farcical solemnities wines were served round to commemorate the day, and the company presented the adored lady with their warmest congratulations. They then dispersed into different apartments, which were finely illuminated, and entertained themselves with dancing, cards, and conversation until nine o'clock, when the doors of another apartment were thrown open in which " sideboards were ranged round and covered in a sumptuous manner with syllabubs, jellies, creams, ices, wines, cakes, and a variety of dry and fresh fruits, particularly grapes and pineapples."

The sequel to the strange flirtation of the reverend doctor has its retributive side. Within a twelvemonth after this glorification, Mrs. Macaulay abdicated her throne at Alfred House, went to Leicester, and there married a younger brother of Dr. Graham, aged twenty-one; whereupon her old admirer lost no time in having the marble monument removed from his church and the vault sold. Mrs. Macaulay's likeness was painted by Gainsborough, and her statue was sculptured by Bacon. According to Boswell, Johnson demonstrated, in his usual trenchant manner, the fallaciousness of the lady's theory of social equality. "Sir," said he to his biographer, "there is one Mrs. Macaulay in this town, a great Republican. One day, when I was at her house, I put on a very grave countenance and said to her, 'Madam, I am now become a convert to your way of thinking. I am convinced that all mankind are on an equal footing; and to give you an unquestionable proof that I am in earnest, here is a very sensible, civil, well-behaved fellow citizen, your footman. I desire that he may be allowed to sit down and dine with us.' I thus showed her the absurdity of the levelling doctrine. Sir, your levellers wish to level down as far as themselves, but they cannot bear levelling up to themselves."

THE SHERIDAN-LINLEY EPISODE.

Incidental mention has been made of Sheridan's association with Bath; but as his career was largely moulded by the brief years he spent in it—years full to him of momentous events—a summary of the facts, as culled chiefly from local records, will now be presented. Richly endowed as he was mentally, Richard Brinsley Sheridan would, under any circumstances, have gained celebrity; but without his residence here and the knowledge he thus gained of the characteristics and foibles of fashionable life, dramatic literature would not have been enriched with comedies as perfect and enjoyable as "The Rivals" and "The School for Scandal," neither would the family have had the captivating beauty which was so conspicuous in the three daughters of Tom Sheridan, an inheritance they received through Miss Linley, "The Maid of Bath." Sheridan's father first visited Bath in 1768-9 in the course of a lecturing tour. He seems to have thought it a desirable place to settle in, and hither he returned with his household in the autumn of 1770. The "Bath Chronicle" of November 22nd advertises "The First Attic Entertainment" at Simpson's Concert Room on the following Saturday; it was to consist of "Reading and Singing: the Reading part by Mr. Sheridan; the Singing by Miss Linley." The entertainment was to be continued on the following Thursday and Saturday; the subscription to be a guinea, for which six tickets were delivered; single tickets, 5s. each. The advertisement is repeated on December 27th, with the following "N.B." added:—"Mr. Sheridan is now ready to receive the commands of such persons as wish to have their children regularly instructed in the art of reading and reciting, and in the Grammatical Knowledge of the English Language. And by the desire of several gentlemen, whose sons have returned home during the holidays, he will immediately receive pupils at his own House in King's-Mead Street, till a proper place shall be fitted up for their accommodation." As Moore in his "Life of Sheridan" states that the family removed to New King Street, which was then nearly finished, it is probable that "the proper place" was found there. Thrice during the month of January the "Attic Entertainment" was given, and as a further inducement to the public they were assured that "particular care will be taken to have the Room made as warm as possible." On every occasion the services of Miss Linley, the youthful siren, whose voice and beauty were the rage of the town, were secured, and this association helped of course to promote the intimacy of the Sheridan and Linley families; but although Sheridan the elder was thus partly responsible for the mutual attachment that ensued between his son Richard and Miss Linley, he was relentless in his refusal to sanction their union.

In the "Chronicle" for October 10, 1771, appears Richard Sheridan's burlesque poem, called "The Ridotto at Bath," which describes the opening of the new Assembly Rooms in "an original Epistle from Timothy Screw, under server to Messrs. K-lff, and F-zw-ter, to his brother Henry, waiter at Almack's." The humour of the composition, the style of which Thackeray carried to such perfection, caused it to be very popular; several editions of it were sold, at the price of one penny. In the meantime Sheridan had become deeply enamoured of Miss Linley, who had been affianced to old Mr. Long, which ended in a rupture and the payment of £3,000 by the aged wooer; she was also persecuted by the attentions of Captain Matthews, a Welshman, and noted card player, and among her other lovers, though she was only sixteen, were Charles Sheridan, the brother of Richard, two baronets, and Nathaniel Brassey Halhed, Sheridan's schoolfellow at Harrow. Sheridan and Halhed wrote in conjunction a farce called "Jupiter," but which neither Garrick nor Foote thought good enough for production. Other literary projects were discussed by them, though their united ages numbered only 38 years. It is singular that Sheridan contrived to conceal his attachment for Miss Linley both from his brother Charles and Halhed. Neither suspected him of being a rival; so guileless was Halhed that when away he commissioned his friend to woo in his behalf, and later on, his father compelling him to accept a writership in the East India Company's service, he wrote a sad farewell to "dearest Eliza" and requested Sheridan to give it any finishing touches his better judgment might suggest. Even Captain Matthews made him the confidant of his hopes, fears, and schemes, not knowing, any more than the other competitors for the young lady's hand, that the trusted friend who appeared insensible to the charms of the much-adored beauty was astutely playing his own game. Later, when Matthews, in a vindictive letter, denounced Sheridan as a "treacherous scoundrel" he had perhaps more justification than for other objurgations he embodied in the same epistle. That Miss Linley was a coquette and gave a certain measure of encouragement to one and another of her suitors is certain; that she had yielded her heart, if unconsciously, to Sheridan is, perhaps, not less certain. When any of her followers on whom she had bestowed alluring smiles were inclined to prove troublesome it was generally to Sheridan she turned to get her out of the difficulty. And, then, how charmingly in verse this "snake in the grass," as he was afterwards regarded by other aspirants, could press his suit let the subjoined lyric tell:—

Dry be that tear, my gentlest love,
  Be hush'd that struggling sigh,
Nor seasons, day, nor fate shall prove
  More fix'd, more true than I.
Hush'd be that sigh, be dry that tear,
Cease boding doubt, cease anxious fear,
    Dry be that tear.

Ask'st thou how long my love will stay,
  When all that's new is past?
How long, ah! Delia, can I say,
  How long my life will last?
Dry be that tear, be hush'd that sigh,
At least I'll love thee till I die.
    Hush'd be that sigh.

And does that thought affect thee too,
  The thought of Sylvia's death,
That he who only breath'd for you,
  Must yield that faithful breath?
Hush'd be that sigh, be dry that tear,
Nor let us lose our heaven here.
    Dry be that tear.

Both the lovers well knew that their respective fathers frowned on their attachment, and came to the conclusion that the way to cut the Gordian knot was to elope. Accordingly, all arrangements being made by Sheridan, they transported themselves to France, where it was given out that they were clandestinely married. The fugitives were pursued by Mr. Linley, who brought back his erring, but, to him, very profitable daughter, for whom he had several engagements pending. Captain Matthews tried also to "improve the occasion" after a fashion, as will be seen from the letter subjoined:—

                        "Bath, Wednesday, April 2, 1772.

    "Mr. Richard Sheridan having attempted, in a letter left behind him for that purpose, to account for his scandalous method of running away from this place by insinuations derogating from my character, and that of a young lady, innocent as far as relates to me and my knowledge; since which he has neither taken notice of the letters, or even informed his own family of the place where he has hid himself;—I can no longer think he deserves the treatment of a gentleman, and therefore shall trouble myself no further about him, than in this public method to post him as a liar and a treacherous scoundrel. And as I am convinced that there have been many malevolent incendiaries concerned in the propagation of this infamous lie, if any of them, unprotected by age, infirmities, or profession, will dare to acknowledge the part they have acted, and affirm to what they have said of me, they may depend on receiving the proper reward of their villany in the most public manner. The world will be candid enough to judge properly (I make no doubt) of any private abuse on this subject for the future, as nobody can defend himself from an accusation he is ignorant of.   "THOMAS MATTHEWS."

When penning this scurrilous epistle, the gallant Captain was probably under the impression that he should neither see nor hear anything more of his hated rival for some time to come. If he did indulge in this comforting reflection he was very soon undeceived. Sheridan, on returning from the Continent in the wake of the Linleys, heard rumours of the above letter, and Matthews being then in London he went to the lodgings of the latter with pistols to avenge the insult. Matthews, however, soothed his irate visitor by assuring him that he misunderstood the purport of the advertisement, wich meant nothing more than an inquiry about his location, put in with the sanction of his family. Finding on reaching Bath the real nature of the libel, Sheridan promptly went back to London, accompanied by his brother Charles, and a duel with swords followed in Hyde Park. Matthews, worsted in the encounter, begged for his life, and signed an apology dictated by his antagonist. This (published in the "Chronicle" of May 7, 1772) was as follows:— "Being convinced that the expressions I made use of to Mr. Sheridan's disadvantage were the effects of passion and misrepresentation, I retract what I have said to that gentleman's disadvantage, and particularly beg his pardon for my advertisement in the 'Bath Chronicle.'   "THOMAS MATTHEWS."

Matthews could ill brook his humiliation, and sought to rehabilitate his character by forcing another duel on Sheridan. Under date July 1 appears the following paragraph:

"This morning, about three o'clock, a second duel was fought with swords between Captain Matthews and Mr. Richard Sheridan, on Kingsdown, near this city, in consequence of their former dispute respecting an amiable young lady, which Mr. M. considered improperly adjusted, Mr. S. having since their first rencontre declared his sentiments respecting Mr. M. in a manner that the former thought required satisfaction. Mr. Sheridan received three or four wounds in his breast and sides, and now lies very ill. Mr. M. was only slightly wounded, and left this city soon after the affair was over." In the next number another paragraph gives the following additional particulars:—"The last affair between Mr. Matthews and Mr. Sheridan, we are assured, was occasioned by Mr. S.'s refusal to sign a paper testifying the spirit and propriety of Mr. M.'s behaviour in their former rencontre. This refusal induced Mr. M. to send him a challenge, which was accepted, and Kingsdown was the place appointed for the decision of their quarrel. After a few passes, both their swords were broken, Mr. S.'s almost to the hilt, who thereupon closed with Mr. Matthews, and they both fell. Mr. M. having then considerably the advantage called on S. to beg his life, which he refused (having in their former duel given M. his life), upon which M. picked up a broken piece of sword, gave S. the wounds of which he last Wednesday lay dangerously ill, and immediately left this city, as before mentioned. The seconds stood by quiet spectators." Immediately after we find: "'Tis with great pleasure we inform our readers that Mr. Sheridan is declared by his surgeon to be out of danger." Captain Matthews spent the rest of his days in Bath, dying in 1820.

When convalescent, Sheridan removed to London, became a member of the Middle Temple, and on the 13th of April, 1773, he was "united in the holy bonds of matrimony to Elizabeth Ann Linley." He was in his 23rd and she in her 20th year. The "Chronicle" thus announces the auspicious union:—"We have the best authority to assure the public that Mr. Richard Sheridan, now a student of law in the Middle Temple, was yesterday married in London to the justly celebrated and admired Miss Linley."

Before leaving Bath, Sheridan (under the "nom de plume" Horatio) gave publicity to two epistles, one of which, as will be seen, is eminently suggestive of the morals of the time:—"The following letters are confidently said to have passed between Lord G——r and the celebrated English syren, Miss L——y. I send them to you for publication, not with any view to increase the volume of literary scandal, which, I am sorry to say, at present needs no assistance, but with the more laudable intent of setting an example for our modern belles, by holding out the character of a young woman who, notwithstanding the solicitations of her profession and the flattering example of higher ranks, has added incorruptible virtue to a number of the most elegant qualifications. Yours, &c., HORATIO."

[Lord G——r to Miss L——y.]

"Adorable Creature,—Permit me to assure you in the most tender and affectionate manner that the united force of your charms and qualifications have made so complete a prisoner of my art (sic) that I dispair of it being set at liberty but through your means. Under this situation I have it ever to lament that the law will not permit me to offer you my hand. Here I cannot resist my fate, but what I can dispose of, my heart and my fortune, are entirely at your devotion, thinking myself the happiest of mankind should either be acceptable. Lady A., who will deliver you this, and who obligingly vouchsafes to be my mediator, will, I flatter myself, urge the sincerity of my heart on this occasion, so as to obtain a permission for me to throw myself at your feet to-morrow evening, in momentary expectation of which I am your devoted admirer, G——r."

[Miss L——y to Lord G——r.]

"My Lord,—Lest my silence should bear the most distant interpretation of listening to your proposals, I condescend to answer your infamous letter. You lament the laws will not permit you to offer me your hand. I lament it, too, my Lord, but on a different principle—to convince your dissipated heart that I have a soul capable of 'refusing' a coronet when the owner is not the object of my affections—'despising' it when the offer of an unworthy possessor. The reception your 'honourable' messenger met with in the execution of her embassy saves me the trouble of replying to the other part of your letter, and (if you have any feeling left) will explain to you the 'baseness' as well as the 'inefficacy' of your design.—L——y."

THE BATHEASTON VASE.

The Batheaston Vase is responsible for the largest quantity of miscellaneous poetry or rhyme ever offered at one shrine. Four volumes of selections from the mass contributed are still extant, and a fifth was ready for publication when the death of one of the originators brought these literary competitions to an end. Looking at the number of aspirants who entered the list, and the time the fêtes lasted, one can imagine the small mountain of manuscript that must have been piled up at this Parnassus. From the examples given in print, it does not appear that anything of a high class was produced. "The poet's eye with fine frenzy rolling" is not apparent in any of the productions; the play of fancy and imagination was, for the most part, weak, but common-place thoughts and reflections are often neatly and prettily expressed with rythmic grace. The amusement, from its novelty, naturally aroused the contempt of matter-of-fact people, and the ridicule of the cynical; but, after all, if it produced nothing that lingers in the memory it stimulated mental effort, and aroused a spirit of emulation in its exercise among the idle votaries of pleasure who here congregated. It compelled them to read as well as to think, and if they did not succeed in gaining the laurel wreath or sprig of myrtle, greater facility in using the pen was acquired, or a better mode of expressing ideas in ordinary correspondence was gained. Apart from these advantages, there can be no doubt that the "re-unions" at Batheaston gave opportunities for enjoyment and social intercourse amidst the charming condition of beautiful grounds and fine surrounding scenery, in comparison with which the ballroom or Pump Room faded in estimation.

The founders of this entertainment were Sir John and Lady Miller, the former an Irish gentleman, the latter an English lady. They built the house known as Batheaston Villa, and laid out the grounds with taste; but having spent more than their resources justified, the young pair betook themselves to the Continent to retrench. In 1770-71, they made a tour of Italy, and in 1776, the letters which Mrs. Miller had sent during her travels to a friend were published anonymously in three volumes under the title of "Letters from Italy, describing the manners, customs, antiquities, paintings, etc., of that country." That the vivacious, if not very critical contents of the work were appreciated is shown by the fact that a second edition, in two volumes, was issued in the following year. In the meantime the Millers had returned to their elegant abode at Batheaston. Among the numerous paintings and works of art which they brought back from Italy, was a beautiful vase, found in 1769 at Frascati, near the spot where is supposed to have stood the Tusculanum of Cicero, "and by its workmanship (says a contemporary writer) seems not unworthy of such an owner." Lady Miller having resolved to naturalise a little Gallic institution called "Bouts Rimés"—rhyming words being given out to be filled up in metre by the competitors—arranged that the contributions should be deposited in this classic urn, which was accordingly raised on a modern pedestal and placed in the bay window of her villa for that purpose. The experiment, started in 1775, was a great success. Indeed, it is surprising to find how many residents and visitors, great lords and fine ladies, authors and wits, contended for the wreaths of myrtle, or the sprigs of bay, given as the reward of distinguished merit. The fact suggests the question whether the talent of versifying must not have been more widely diffused or cultivated than it is at the present day, more particularly as the pastime flourished until the death of Lady Miller in 1781. The Rev. R. Graves, rector of Claverton, records that one morning he counted over fifty carriages of visitants drawn up on the road below Batheaston Villa, and that he was once present when there were four duchesses in the company, viz., Beaufort, Cumberland, Northumberland, and Ancaster.

Fanny Burney, who was familiar with the proceedings at the villa, describes the attraction they had for the "elite" of Bath, despite the ridicule with which they were treated in some quarters. "Nothing in Bath," she writes, "is more tonish than to visit Lady Miller, who is extremely curious in her company, admitting few persons who are not of rank and fame, and excluding of those who are not all persons of character very unblemished." Her husband, however, had more conceit than refinement. Lady Spencer and Lady Georgina Spencer were present on one occasion when Sir John Miller read a poetical offering addressed to Lady Georgina, the flattery of which was so gross as to distress the lady, who blushed and looked down in the utmost confusion. Sir John hinted that the author deserved the prize for having chosen so fine a subject, which was agreed to, whereupon he avowed himself the author, adding "and now I must read them over again." Lady Georgina curtsied, and hastily retired with Lady Spencer.

Among the competitors whose pieces were chosen for publication were the Duchess of Northumberland, the Marquis of Carmarthen, Earl Temple, the Earl of Carlisle, Lord Palmerston, Lord Nugent, Admiral Keppel, Sir John and Lady Miller, Rev. R. Graves, Dr. Whatley, G. Pitt, John Jeffreys Pratt, Anstey, Schomberg, Drax, Garrick, Skrine of Warleigh, Melmoth, Hon. Mrs. Greville, Miss Bowdler, Miss Seward, etc. The profits on the sale of the poems were devoted to the Pauper Charity (the predecessor of the Royal United Hospital), of which Sir John Miller was for some time president. The report for 1776 acknowledges the receipt of £10 as a first instalment from this source.

The gatherings were held weekly, the order of procedure observed at them being as follows:—The subjects were given out at one meeting, and at the next, as the company assembled, each competitor deposited his or her composition in the vase, after which a lady, chosen by Lady Miller, drew out the papers promiscuously, and gave them in succession to the gentlemen present to read aloud. At the conclusion, a select committee were named to act as censors, who chose three poems which were deemed most worthy of the myrtle wreaths, and the authors, or their proxies, having read them a second time to the company, were presented with the evergreen rewards. These, if conferred on a gentleman, were ordinarily handed to some lady, who wore them at the Rooms on the ensuing ball night. "Everything," says a wearer of the bays, "is conducted with the utmost propriety and elegance. Those who have no taste for poetry may amuse themselves with some capital paintings, brought from Rome, with which the apartments are adorned. Those who have no taste for either may at least entertain themselves with the sight of many a blooming beauty, and by the generous bounty of the lady of the house, with chocolate, jellies, biscuits, and macaroons." To maintain harmony and good feeling all discussion of party and opinion, and all tendency to personality, or the violation of the sanctities of society, were rigidly forbidden and discouraged. A curious circumstance illustrates the prudence of these rules. At one of the balls, Lady Brownlow North, wife of the Bishop of Winchester, refused to comply with the stated rules as to full dress, and paid no heed to the remonstrance of the M.C. A great commotion was caused, and as the company in general took the part of their chosen monarch, her ladyship found her defiance attended with disagreeable consequences. At the next assembly at Batheaston Villa, Mr. Cradock had some verses handed to him to read. He saw that the subject chosen by the writer was Lady North's escapade, and, ceasing to read, placed the paper in his pocket. A nobleman present wanted to hear the lines. Rather than comply Mr. Cradock left the room, and burnt the satire. Wilkes, who was then in Bath, was believed to have been its author. (Lady North, née Miss Bannister, was one of the leaders of fashionable society, and her extravagance absorbed no small portion of the rich emoluments of Winchester See).

Whatever may have been the merits of Lady Miller's literary scheme, it was deemed too frivolous to commend itself to Johnson, as we know from Boswell. Horace Walpole also found it a congenial subject for his skittish pen. In a fanciful description, he states that "a Roman vase, decked with pink ribbons and myrtles, receives the poetry, which is drawn out every festival; six judges of these Olympic games retire and select the brightest compositions, which the respective successful acknowledge, kneel to Mrs. Calliope Miller, kiss her fair hand, and are crowned with myrtle—with, I don't know what. You may think this a fiction or exaggeration. Be dumb, ye unbelievers! The collection is printed, published. Yes, on my faith, there are bout rimées on a buttered muffin made by her Grace the Duchess of Northumberland; receipts to make them by Corydon the venerable, alias George Pitt; others, very pretty, by Lord Palmerston; some by Lord Carlisle; many by Mrs. Miller herself, that have no fault, but wanting metre; an immortality promised to her without end or measure. In short, since folly, which never ripens to madness, but in this hot climate, ran distracted, there never was anything so entertaining or so dull." Of course the foregoing is a highly-coloured picture, with a dark background of malevolence, which Walpole was fond of painting for the gratification of his correspondents. Disappointment may have been an incentive in this case; his own contributions to the vase not having secured for him the myrtle crown for which he affected the same scorn as the fox did for the grapes. The many enjoyed the festival as an innocent, healthy, and intellectual pastime; but it was an exotic not suited to the soil of this country, and so it perished. Lady Miller died on June 24th, 1781, in her 41st year, and was buried in the Bath Abbey, where on the north side of the altar a monument in statuary marble, by Bacon, perpetuates her memory.

Lady Miller's House

CHAPTER IX.

SCIENCE AND ART.

HE chief local contribution to the science of the century was the improvement in the telescope effected by William Herschel, and the discovery by him of the planet Uranus. He came to Bath in 1766, thinking to find profitable scope for his talents as a musician; nor was he disappointed; but his favourite pursuit was astronomy, to which he devoted all his spare time. He used to retire to bed with a basin of milk, or glass of water, at his side, and pore over Ferguson's "Astronomy" and other kindred works until he went to sleep, while his first thought on rising was how to obtain instruments for viewing those celestial objects of which he had been reading. His earliest venture was to hire a two-and-a-half foot telescope, which served not only for viewing the heavens, but for learning its method of construction. Having mastered these particulars, he inquired in London the price of a reflecting mirror for a five or six foot telescope. The reply was that there were not any of so large a size, but one could be made at a price named, which Herschel's means would not allow him to face. Besides, at this juncture he purchased the tools, patterns, and unfinished mirrors of a Quaker optician in Bath, and though none of the mirrors was of the dimensions required, the stock he had acquired encouraged him to set about making a telescope for himself. He was then residing at 7, New King Street, and his sister Caroline, who was his house-keeper and assistant, records the consequence of this resolve. Writing in the summer of 1773, she says: "I saw, to my sorrow, almost every room in the house turned into a workshop—a cabinet-maker making tubes and stands of all descriptions in a handsomely-furnished drawing room; my brother Alexander putting up a huge turning machine (brought from Bristol) in a bedroom for turning patterns, grinding glasses, and turning eye-pieces." In the following year the Herschels removed to a house near the Walcot Turnpike, in the rear of which was a grass plot, whereon a twenty foot telescope was erected, and a twelve foot mirror prepared for it. So concentrated was William on his work that his sister had occasionally to put victuals into his mouth to keep him alive; and in one instance, when anxious to finish a seven foot mirror, he did not take his hands from it for sixteen hours.

Herschel was not satisfied with his Walcot location and returned to New King Street, this time to No. 19, a larger house with a garden behind it extending to the river. Here he placed his twenty foot telescope and applied himself to perfecting his mirrors. He had set his mind on making one for a thirty foot reflector, and a furnace for it was built in a room on a level with the garden. The mirror had to be cast in a mould prepared from horse dung, an immense quantity of which was pounded in a mortar, and passed through a sieve. It was an endless piece of work, and gave Caroline many an hour's exercise, as well as Alexander, and even Sir William Watson, when calling to see Herschel, would take the pestal from Caroline and pound away vigorously at the manure. On the day set apart for casting, the metal in the furnace unfortunately began to leak at the moment when ready for pouring, and both the Herschels, with the workmen, were obliged to rush out of the door, for the stone floor was blown about in all directions, and as high as the ceiling. William fell, exhausted with the heat and exertion, on a heap of brickbats. Nothing daunted, he at once made another attempt, which proved entirely successful; the much-longed for mirror was produced, and, being adjusted to the tube, the persevering astronomer was rewarded with the discovery of Uranus and with fame that became world-wide in extent. On the establishment of the Bath Philosophical Society in 1781, he contributed papers to its proceedings; but in the following year, being appointed Astronomer Royal at Windsor, he severed his connection with Bath.

Another scientific genius, whose reputation is as lustrous as ever, was William Smith, the "Father of English Geology." It was while engaged in making surveys for the Kennet and Avon Canal in the last decade of the century that he accumulated the facts establishing the truth of a generalisation he had evolved, that the same strata were always found in the same order of superposition, and contained the same fossils. His discovery unlocked the secret of stratification as much as the Rosetta stone supplied the key to the Egyptian hieroglyphics.

The application of science to agriculture was greatly stimulated by the establishment, in 1777, of the Bath and West of England Agricultural Society. Its programme, however, was not limited to this one branch; it not only aimed at making the land more productive by better tillage, and improving stock by more care in breeding, but it sought to develope manufactures by encouraging invention, and to foster art by extending to it a helping hand. So far as the farming interest was concerned, it needed enlightenment. The methods employed were primitive, particularly in the Western Counties. The implements in use were limited in number and simple in construction. The agriculturist was guided by custom, or by such modifications of it as observation and experience might suggest. He ploughed and he sowed as his ancestors had done from time immemorial; knowing little of the advantages of manures, his crops were, as a rule, small and the quality inferior. More likely than not he threshed his corn after combing the sheaves with a hand comb, and cutting off the heads with a common knife. His land for a great part of the year lay fallow because he did not understand the rotation of crops. With swede turnips or mangold wurtzel he had not become acquainted, and his cattle were raw-boned and ungainly, for not only was the food inadequate, but the plan of cross-breeding was unknown. In some other parts of England more advancement was shown, particularly in Norfolk, where Mr. Edmund Rack was born. A man of literary tastes, he came to Bath, and was a well-known figure among the authors here resident. He was struck with the rude condition of agriculture, and to remedy it, proposed the founding of a society in the city "for the encouragement of Agriculture, Planting, Manufactures, Commerce and the Fine Arts." The proposal was favourably received, the Society was organised, the Earl of Ilchester being the first president, Mr. Rack the secretary, and operations were commenced.

The first subject to engage the Society's attention was how to grow corn on the best and cheapest method, with due regard to economy of seed and sufficiency of yield. A prize of £10 for "setting" ten acres on the Norfolk system, was awarded to Mr. Thomas Bethell, of Weston. An experimental farm was also established, where trials of new machinery and of planting were conducted by earnest and unprejudiced men. Blancher's drill plough was so tested and certified to "deliver the grain with great exactness and regularity." The Society examined and expressed its approval of Winlaw's mill for separating the ears of corn in place of threshing. Indeed, every fresh contrivance or change likely to benefit the tillers of the soil was promptly welcomed and put to the proof. But the bucolic mind was slow to adopt new ideas; vigorous attempts were made to introduce the Norfolk and other improved ploughs calculated not only to expedite the work, but to render it less expensive, yet it was a subject of complaint that not one farmer in five hundred used them, though many received daily ocular demonstration of the inferiority of their own ploughs. Premiums were accordingly offered to ploughmen who adopted and properly used the Norfolk implement, and this led to its more general employment. Many other inventions were patronised by the Society, some of which secured almost immediate success to the great advantage of the inventors and the encouragement of the ingenious. New kinds of drill, dibbling, threshing, and winnowing machines are mentioned, while one, patented by Mr. Lazarus Cohen, of Exeter, in 1796, is described as "a curious machine for reaping and mowing, by the use of which one man, with great facility, can do the work of three;" but though deemed a practical invention, it was not perfect enough to be entitled to the Society's premiums. It would be interesting to know the fate of this labour-saving contrivance, which fancy suggests may have, in a primitive form, embodied the principles of the modern reaping and mowing machines. Further, in memory of Francis, Duke of Bedford, who was president of the Society at the close of the century, a gold medallion, value twenty guineas, was offered annually as a prize for the greatest improvement in any subject connected with agriculture. So rapid was its progress that the Society's meetings in December were always attended by many noblemen, gentlemen farmers, manufacturers and tradespeople, who took a lively interest in its proceedings. The "Transactions of the Society" were published annually, and these volumes contained a large amount of information by competent writers, invaluable to the practical farmer. A library was also formed, including the best books on agriculture, which were at the service of the subscribers, and in time it grew to about one thousand volumes.

Neither was the Society unmindful of its self-imposed obligation with reference to Art. The Bedfordian medal was awarded to Chantrey, the sculptor, and to Miss Fanshaw, who submitted the best design for the die. Nollekens was commissioned to execute a bust of the Duke, and a like compliment was paid to Sir Benjamin Hobhouse, who was president for several years, the artist in this case being Chantry. Both busts are at the Bath Royal Literary Institution. Artists of merit were also aided by bringing their works under the notice of the affluent, and in one case by a very interesting purchase. In the middle of the last century there were two local painters of the name of Robins, father and son. The former produced in oil one or more large views of Bath, which were engraved, some of the prints being still in the hands of collectors. The son, who, like the father, taught painting and drawing, devoted himself more especially to copying flowers, butterflies and insects. He put together two folio volumes of these subjects, which are a marvel of exquisite colouring and tasteful grouping. The butterflies, moths, and insects are true to nature, both in size and tint, and associated with appropriate flowers, grasses, etc. On the page facing each group the names of the insects and botanical specimens are written and methodically arranged. The beautiful tomes, though, by the way, they have nothing better than canvas covers, ought, one would suppose, to have made the artist famous. Yet his name is nowhere recorded, except in newspaper files, where he advertised himself as a teacher of drawing and painting. That his work is remembered at all is due to the action of the Society. Robins, in his old age, was reduced to a state of indigence from the lack of pupils, and to relieve his necessities he asked the Society to purchase the two volumes, which had been valued at one hundred guineas. Such a sum could not be spared from the funds, and the artist then agreed to take whatever the Society could afford. Twenty guineas were offered, with a condition that the work could be redeemed within two years. The money was accepted, but the condition was not complied with, and the two volumes, with their couple of hundred of choice paintings, thus became the property of the Society, and by it they have been carefully preserved. From this brief summary of its proceedings in its earlier years, it will be seen how beneficent was the influence the Society exercised, especially in bringing science to bear on agricultural pursuits, thereby helping the farmer out of his fossilised habits, enriching the country, and cheapening the food of the people.

Independently of the Society, Art flourished in Bath, as is shown by the number of its professors who flocked to it in the hope of finding it a haven of prosperity. One of the earliest to obtain distinction was Van Diest. He was the son of Adrian van Diest, a Dutchman, who came to England, and was employed by Lord Bath during the last quarter of the seventeenth century to paint several views of ruins in the West of England. On his death in 1704, his son settled in Bath, and in the course of his professional career, he painted the members of the Corporation at the request and expense of Marshal Wade (ante Wade), whose likeness he also limned. The latter, as well as several of the former, still hang in the Guildhall. Another of his works, a portrait of Mr. Dunkerley, who was for some time the head of English Freemasonry, is preserved at the Royal Cumberland Lodge. The altar piece in Allen's private chapel at Prior Park, and a three-quarter length portrait of Wade, for the same patron, were also painted by him. Towards the end of his career he was known as Vandyke, under which name he had a studio in Broad Street. Falling on evil days, he was befriended by Dr. Oliver, at whose house, Trevarno, Bathford, he was a frequent guest.

Van Diest seems to have been overshadowed by a later contemporary, William Hoare, who, having married a Bath lady, took up his abode here about 1739. He was not only a skilful artist, but a man of scholarly tastes, with easy, urbane manners, which made him a favourite in society. His studio was resorted to by all that could boast the attractions of either beauty, fashion, or intellectual eminence. The number of his sitters was so great that he had scarcely any intervals of leisure; yet he was painstaking under the greatest pressure, always striving to do his best. He was a conscientious, if not a great painter. Finding there was a demand for pictures in crayons, he mastered the technique of the style, and that so well that his crayon drawings are thought more highly of now than those he produced in oil. Examples of both kinds may be seen at the Guildhall and at the Mineral Water Hospital. When the Royal Academy was founded, Hoare was chosen one of the original members, his diploma being the last signed by the King. He was a regular contributor to its exhibitions up to 1779, his works being chiefly in crayons, some of which have found a resting place in the National Portrait Gallery. He died in 1792 in his 85th year.

When a greater genius, Thomas Lawrence, arrived in Bath as a boy, marvellous in the use of the brush, he found in Mr. Hoare a friendly guide and counsellor. In a letter to his mother, Lawrence tells her that he gave him all the help in his power. "I have (he adds) a very valuable receipt from him, which was made use of by Mengo, the Spanish Raphael." He was then twelve years of age, and lodging at 2, Alfred Street, whence he issued a prospectus in March, 1782, announcing that in the following May would be published " a very striking and approved likeness of Mrs. Siddons in the character of Zara in ' The Mourning Bride,' drawn from life by Master Lawrence. Five shillings to be paid down and two shillings and sixpence on delivering the impression." The picture was an oval one, twelve by ten. Another of his compositions was " Peter denying Christ," and he also, while here, copied the figure of our Saviour from Raphael's picture of " The Transfiguration."

In the meantime Gainsborough, who had established himself in the Circus, was, from his pre-eminent talent, patronised by the rank and fashion of the time, and painting for his amusement the players and musicians with whom he delighted to associate. Garrick, Foote, Edwin, Quin, and John Henderson, Abel, the viol-di-gamba player, Giardini, the violinist, Fischer, the hautboy player, were thus favoured ; while among his other sitters were the Duke of Argyll, Lord Clare, Lady Grosvenor, Lady Sussex, Lord and Lady Ligonier, Lord Camden, and General Honywood. At the same time he produced the beautiful landscape, " The Return from Harvest," and the fine portrait of Orpin, the old parish clerk of Bradford-on-Avon. Indeed, it is now held that the great artist reached the highest point of excellence during his Bath period.

A few years later, and Thomas and Benjamin Barker were emerging into fame. The talent for drawing the former manifested, fortunately attracted the attention of Mr. Spackman, a Bath coachbuilder, through whose aid he made some good copies from the Dutch masters, and, continuing to improve, the same friend made him a liberal allowance to go to Italy. On his return he exhibited at the Royal Academy and continued to do so occasionally for some years. His " Woodman," " Old Tom," " The Gipsy " and other rustic groups were very popular and were reproduced on china, pottery, and even textile fabrics. His brother Benjamin painted both in oil and water colour, but though he was not without taste and feeling, he met with little encouragement in his profession. He died in 1838, and Thomas in 1847.

Among the less distinguished artists was George James, A.R.A., who painted the portrait of Richard Tyson, one of the M.C.'s, now in the Assembly Rooms. He was a noted bon vivant and comic singer. Having gone to Boulogne, he was thrown into prison by Robespierre, and the hardships he endured shortened his days. Another was James Vertue, who, in addition to portraits, drew the interior of the Abbey, which his brother George engraved. Joseph Wright, (" Wright, of Derby ") spent some time in Bath as a portrait and landscape painter ; but not receiving the encouragement he expected, he returned to his native town. Thomas Beach, a pupil of Sir Joshua Reynolds, commenced his professional career here, and sent portraits to the Incorporated Society of Artists ; he painted, in 1787, Mrs. Siddons and her brother in the " dagger scene " in " Macbeth," which was forwarded to the Royal Academy. John Hodges Benwell executed a few small oval drawings in water-colour which he united with crayon in a manner peculiar to himself —a method much prized at the time. Thomas Bonner, an engraver of repute, engraved the numerous pictures for Collinson's " History of Somerset." William Henry Bunbury, son of the Rev. Sir W. H. Bunbury, Bart., of Mildenhall, Suffolk, was a deft amateur with the pencil, and became celebrated as a caricaturist, among his productions being " The Long Minuet at Bath." Robert Edge Pine, history and portrait painter, practised in Bath from 1771 to 1779. His " Surrender of Calais," the figures in which are life-size, was placed in the Town Hall, Newbury; it was 13½ feet long and 9 feet high, and being deemed too large for the Hall was put up for a raffle. His portrait of Garrick is highly esteemed. He went to America—and no wonder after the raffle—taking with him several of his historical pictures, which were unfortunately destroyed by fire. He died at Philadelphia in 1790. His brother Simon was a miniature painter of high standing, and sent portraits to the Royal Academy in 1772, the year in which he died at Bath. The unfortunate Major John André was an artist of no mean ability; a half-length miniature which he painted of himself was engraved by Sherwin, and a bold landscape etched by him is still preserved. Samuel Collins, who was born in Bristol about the middle of the century, spent several years in Bath painting miniatures on ivory and enamel. His pupils included the more celebrated Ozias Humphrey, R.A., who, on Collins removing to Dublin, succeeded to his connection; later on Humphrey went to India, and many of the native princes and other distinguished persons sat to him; on his return he tried his hand at crayons, and was so successful that he was appointed portrait painter to the King. Another miniaturist who established himself in Bath was Lewis Vaslet; he was an occasional exhibitor at the Royal Academy. Thomas Worlidge, a portrait painter, practised first in Bath in the reign of George II., drawing, with pencil on vellum or Indian ink, portraits of a miniature size; he was a good draughtsman, and etched with much ability after the manner of Rembrandt; he married a young wife, a local beauty, the daughter of a toyman, who assisted him in his profession, and became celebrated for her skill in worsted work. Her husband was careless and improvident; when in want of a dinner in his early days he luckily found half-a-guinea, and instead of spending it on beefsteaks and a pair of shoes, as his wife desired, he indulged in a feast of early green peas. William Watts, an engraver, published twelve views of the city and its environs, which are considered fine specimens of line engraving. John Beauvais, a French miniature painter, practised with success in Bath in the latter half of the century. John Taylor, who was born in Bath in 1745, studied in London and obtained note as a landscape painter. He resided afterwards in the Circus, where he continued to paint and to write poetry as well, which secured for him the title of "artist and poet." In April, 1780, it was announced that "two highly-finished rich engravings from landscapes and ruins, painted by John Taylor, esq., of the Circus," are to be issued to subscribers. Thomas Hickey, a portrait painter of high standing, painted two portraits of Mrs. Abingdon, a full-length of Dr. Warner, and the head of Mr. W. Brereton, M.C. Mrs. Patience Wright, an American lady, a celebrated modeller in wax, had rooms in the Abbey Churchyard, where she modelled many portraits. Abraham Daniel, a miniature painter, whose works are highly esteemed, lived and died in Bath. — Dusign, who was a member of a local family, practised here as a portrait painter, after being a pupil of Sir Joshua Reynolds. He died prematurely from consumption at Rome. Charles Gill was the son of a Bath pastrycook, who placed him under Sir Joshua Reynolds, and on two occasions he exhibited portraits at the Royal Academy. J. Hassell, draughtsman and engraver, drew and engraved sixteen tinted aqua-tints for a "Picturesque Guide to Bath and Bristol." William Hibbart practised chiefly in Bath as a portrait painter about the middle of the century, etching heads also after the style of Worlidge. Prince Hoare, son of the William Hoare (already noticed), was born in Bath, and was trained to his father's profession, painting, among other subjects, a portrait of Sir Thomas Lawrence when a boy; but not being satisfied with his success he devoted himself to literature. William Armfield Hobday, miniature and portrait painter, was a yearly visitor to Bath and found much employment. Charles Sherriff, who was deaf and dumb, practised as a miniature painter here, among his sitters being Mrs. Siddons, who stated that his miniature of her was more successful than that of any other artist. John Robinson, portrait painter, was born in Bath in 1715, and after being a pupil of Vanderbank became distinguished in his profession. Thomas Redmond, miniature painter, settled in Bath in 1764, whence he sent crayons to the Royal Academy from 1775 to 1779. Hugh O'Neil, an architectural draughtsman, taught drawing in Bath, and made many carefully finished drawings. Samuel Lysons, amateur, was skilful alike as an artist and an antiquarian, of which his drawings of local Roman remains furnish proof. Henry Leake, portrait painter, was a son of the eminent bookseller of that name. He was a pupil of William Hoare, and after exhibiting some portraits in London he went to India, where he died. J. Keenan, another portrait painter, was residing in Bath in 1792, in which year he was an exhibitor at the Royal Academy. Charles Jagger, a miniature painter of marked ability, executed a portrait of the Duke of Clarence, which was engraved. He died in Bath after two days' illness at the age of 57. Thomas Walmesley, a landscape painter of Irish extraction, settled in Bath towards the close of his life.

The foregoing list will show that the city was a favourite abode of the brothers of the brush, who, as a rule, fared well under the patronage they received. In 1783, the principal resident artists established " a school or academy, for study from antique statues and the living model, upon such a generous plan as to extend its use, not only to every class of artists, but also to the instruction of youth." The institution was a forerunner of the modern School of Art.

AN OLD BATH WATCHMAN.

CHAPTER X.

MUSIC AND THE DRAMA.

O doubt the culture of music in Bath was largely promoted by the permanent employment of a band at the Pump Room, composed as it was of skilled musicians. It was further aided by the fact that the organist at the Abbey for a third of the century was not only a distinguished performer, but a composer of repute as well, viz., Thomas Chilcott. He was appointed to the post in 1733, and held it until the 24th of November, 1766, when he died. Among his compositions are twelve songs, the words by Shakspere, Marlowe, Anacreon, and Euripides, and six concertos, which were dedicated to Lady Elizabeth Bathurst. He was a keen musical critic, and a favourite in society from his talents both as an instrumentalist and vocalist. At the Duke of Beaufort's as well as at Earl Bathurst's seat, and at the residences of the country gentry, he was a welcome visitor. It is an interesting fact that he was the means of inspiring the musical aspirations of Thomas Linley, who was the son of a carpenter and was born at Badminton in 1733. Some ten years later the family removed to Bath, where the extensive building then in progress gave ample employment to the father, and the son (according to Mr. E. Green, who with indefatigable zeal has tracked the fortunes of the Linleys), became an errand boy to Chilcott, who soon detected the lad's musical ability, and, with the parent's consent, took him as an articled pupil for five years. Before his apprenticeship expired, the youth was a master of the theory and practice of music. He married at the age of 19, and set up for himself as a teacher of singing and the violin, in both of which branches he won a high reputation. Hitherto concerts had been given at Harrison's or Wiltshire's rooms, where the most noted singers from London and the provinces were to be heard. They were sporadic undertakings. Linley became the "entrepreneur" of these speculations, and conducted them upon spirited and systematic lines. He was fortunately aided by members of his young family, several of whom inherited their sire's genius. Dr. Burney, who visited them at their residence in Pierrepont Street, described them as a "nest of Linnets"; indeed it is rare that talent of such conspicuous merit is exhibited by so many members of one family. The eldest son, Thomas, showed, when a child, marked skill on the violin, and at the early age of seven was taken as a pupil for five years by Dr. Boyce. When eight years of age he performed in public, and at twelve he composed six solos for the violin. He subsequently went to Florence, where he made the acquaintance of Mozart, who became warmly attached to him, and, so perfect was his mastery of the instrument, that he ranked as one of the finest violin players in Europe. In 1773 he was leader of the orchestra and solo player at his father's concerts at the Pump Room and at the Drury Lane oratorios. His sisters, Elizabeth and Maria, were accomplished vocalists when they had just entered their teens. The former was the Jenny Lind of her day. In both the exquisite quality of their voices was combined with unerring sympathetic feeling, which enabled them to give artistic expression to compositions lively or pathetic, as well as the grand Handelian solos. Their personal beauty was as fascinating as their voice. Elizabeth, as is well known, became the wife of Richard Brinsley Sheridan; her sister was early removed by death. A touching incident marked her closing days; she rose in her bed and sang with such sweetness and fervour the beautiful air "I know that my Redeemer liveth" that the relatives gathered around were unable to restrain their emotion, knowing as they did too well that it would be her last song on earth ere joining the celestial choir. Another sister, Mary, inherited something of the same musical ability. In 1771, when twelve years of age, she appeared at the Three Choirs Festival at Hereford, and the following year at Gloucester. She married Richard Tickell in 1789, and died seven years later.

The income Linley derived from his profession and the performances of his sons and daughters enabled him to remove to the Royal Crescent, and to buy an estate at Didmarton, Gloucestershire. The elopement of his favourite daughter, Elizabeth, with Sheridan, was a misfortune to himself, however great the lustre the brilliant genius of his son-in-law may have shed upon the family.

No professional man perhaps enjoyed a more enviable position than did Thomas Linley during the brief years that he had his gifted children with him. Unhappily, financial losses and domestic bereavements threw a gloom over his later years. On leaving Bath he became joint manager with Stanley, of the Drury Lane oratorios, and, in conjunction with his son, he composed the music of Sheridan's comic opera "The Duenna." He was, however, involved in his son-in-law's speculations at Drury Lane Theatre; and his children also fell from him like autumn leaves. Of the twelve born only three outlived him, one of whom, Ozias Thurston, was a musical amateur and organist and became a minor canon of Norwich Cathedral, Vicar of Stoke Holy Cross, Norfolk, and afterwards organist at Dulwich College. The most crushing blow was the sad death of his eldest son. He was visiting the Duke of Ancaster, at Grimsthorpe, Lincolnshire, when, owing to the capsizing of a boat, he was drowned. The accident happened on the 5th August, 1778. The body, when recovered, was buried in the family vault of the Ancasters. The calamities which fell upon the elder Linley undermined his health. He had an attack of brain fever after the death of his son, and from that time he never recovered his strength. He died suddenly at his house, Southampton Street, Covent Garden, on the 19th November, 1795, and was buried in Wells Cathedral, where other members of his family are interred, as tablets in the cloisters record.

Linley's successor was his rival, William Herschel, whose fame as an astronomer has eclipsed his reputation as a musician. He was a thorough master of the laws of harmony, and his instrumental ability placed him in the front rank of executants. To his devotion to the art he was indebted for the means of prosecuting his scientific researches. A Hanoverian by birth, he came to Bath in the early part of 1767, and on the opening of the Octagon Chapel in October of that year he was appointed the organist. On the 28th and 29th of the month he gave a performance of "The Messiah" in the chapel, the choruses being rendered by gentlemen of the choirs of Oxford, Salisbury, and Gloucester, the instrumentalists being from this city, Oxford, and London. Between the second and third parts an organ concerto was given by Mr. Herschel, "first violin and manager of the performance."

On the evening of the first day a concert, under his auspices, took place at Wiltshire's Rooms. Henceforward, the young Hanoverian was actively engaged in giving sacred performances at the Octagon Chapel, conducting oratorios at the Rooms and Theatre, and concerts at Spring Gardens and elsewhere. Handel was very popular to judge by the regularity with which "The Messiah," "Judas Maccabæus," "Samson," and "Acis and Galatea" were produced both by Herschel and Linley. The former was a member of the latter's band, until a rupture occurred owing to Herschel considering himself discourteously treated by Linley. With the retirement of Linley, Herschel's duties became more onerous. Laborious as they were in the chapel, the concert-room, and the theatre, he found time to give some thirty lessons a week to pupils, and to compose concertos for the organ, as well as glees, catches, etc. Besides being an organist and a violinist, he played on the harpsicord, the hautboy, the oboe, and the violoncello, the latter being the special instrument of his brother, Alexander, who was a member of the band and married a Bath lady. Caroline Herschel, whose voice and style of singing were much commended, greatly assisted her brother; she sang at the Octagon Chapel, took part in the oratorios, instructed the treble singers, copied the scores of "The Messiah" and "Judas Maccabæus" for an orchestra of nearly a hundred performers, and the vocal parts of "Samson." With the growth of Herschel's renown as an astronomer, his musical avocations ceased. He opened a new organ at St. James's Church in May, 1782, when two performances of "The Messiah" were given, and on Whitsunday following both brother and sister played and sang for the last time in public at Margaret's Chapel, the anthem selected for the day being one of Herschel's own composing. Numerous as were his productions, only one, "The Echo" catch, ever appeared in print.

Happily at this juncture there was another musician who was in every respect qualified to assume Herschel's mantle. This was Venanzio Rauzzini, a handsome Italian, with a fine tenor voice and an enthusiastic devotee of his profession. He was born at Rome in 1747, and having been trained for the operatic stage he came to London, and had a highly successful career. He wrote among other pieces an opera called "La Vestale," which, when produced at the King's Theatre, proved a total failure. He left London, and resided in Bath, where he spent the remainder of his days. His local career commenced in 1777, he having La Motte as his coadjutor—a partnership that did not last very long. For some twenty years he produced, season after season, a succession of high-class concerts, for which he procured the best soloists in the country, and drilled both choir and band to a pitch that satisfied approximately his exacting taste. To secure the excellence he desired he was not always careful to sit down and count the cost, or formed too sanguine an estimate of public support. His enterprise was, in consequence, requited sometimes with financial failure, which was increased by his easy, trusting nature. From a statement published by him in November, 1789, it appears that the expenditure on the previous winter concerts was £521 4s. 9d., and the total receipts £453 6s. 6d., leaving a deficit of £67 18s. 3d. He further explained that the principal performers received no emolument beyond the receipts for their benefits. To give an idea of his operations it may be mentioned that in one week he gave three evening concerts at the Assembly Rooms, Handel's compositions being prominent in the programme, and a morning performance of "The Messiah" at the Abbey; the band alone numbered over 150 performers. Nor was the full tide of harmony thus poured forth at all exceptional. During the summer months Rauzzini was wont to give "elegant serenades" at his villa at Widcombe (Woodbine Cottage) to his friends, when the leading vocalists of the day could be heard. Fortunately he was acknowledged to be one of the best cultivators of the voice in England, and his reputation gained him many pupils, among the more distinguished of whom were Braham, Mrs. Billington, Selina Storace, and Incledon. The tablet to his memory in the Abbey was erected by Braham and Storace. Rauzzini produced several sonatas, arias, and songs, as well as operas.

During his directorship came first into prominence a name that for long years after occupied a foremost position in the musical world—Loder. A skilful violin player at the concerts was John Loder the elder, who died in the prime of life at Weymouth in 1795, leaving his family in straitened circumstances, to relieve whom a benefit concert was held at the Assembly Rooms, at which Braham and other distinguished singers gave their services. One noteworthy item was a concerto on the violin, played by "Master J. Loder, eight years of age." The juvenile performer was the John David Loder, who was for many years at the head of his profession in Bath, conducting the band at the theatre, giving concerts in conjunction with Sir George Smart, and leader of the Three Choirs Festival at Gloucester, Worcester, and Hereford, and whose sons, Edward James, John Fawcett, and William were all eminent musicians. Miss Loder, the sister of John David, was a popular singer at the concerts in the last decade of the century.

The taste for music was also cultivated by the existence of one or two glee and catch clubs, the principal of which was long fostered by Dr. Harington, who was well versed in the science and skilled in the composition of music, of which the numerous glees, catches, and duets he wrote attest, as well as his sacred piece, "Eloi." The latter was thought so highly of that it was customary to sing it between the first and second parts of "The Messiah," as well as every Good Friday at the Abbey and other churches. In conjunction with the Rev. J. Bowen he, in 1795, founded the Bath Harmonic Society, the doctor being president and his colleague vice-president. Among the numerous compositions the former wrote for the Society was the following glee for three voices, which was always sung after the toast of "The King":—

Here's to Rex, Lex, and Pontifex,
A toast no loyal heart rejects:
The King in safety all protects,

The Church to future bliss directs;
But knaves who plot the State to vex,
May Law provide for all their necks!
Amen.

THE DRAMA.

The drama in Bath shared in the general progress the city enjoyed, but had its ups and downs here as elsewhere. Its early beginnings were mean, though particulars are wanting to reveal what they were exactly like. That there was a theatre here when the century opened is shown by the performances that took place on the visit of Queen Anne in 1702. In the Royal train came the Drury Lane Company, who made their debût in a stuffy little theatre which was graced by the presence of her Majesty and her Consort, Prince George of Denmark. It was on this occasion that Nance Oldfield, who became a most charming and popular actress, and one of the progenitors of the Cadogan family, established her reputation by the easy, coy way she played the part of Leonora in "Sir Courtly Nice." The house, room, or, as some say, "stable," where entertainments of this kind took place was situated close to the town walls on the northern side; the eastern wing of the Royal Mineral Water Hospital now stands on the site. After the Royal visit the citizens were ashamed of their Theatre, just as they realised other shortcomings—the lack of a Pump Room to wit—and they resolved to improve it, or, acquiesced in others doing it, as the funds are said to have been provided by people of the highest rank, whose liberality was rewarded by their arms being emblazoned in the interior of the new building. The latter replaced the old structure, which was demolished, and as its successor cost some thirteen hundred pounds, it must have been of fairly good dimensions. In it companies of strolling players of the better class strutted and fretted their hour on the stage; while others of inferior degree had to be content with any large room the public houses could supply. All went well until 1736, when the Act of Parliament was passed for the suppression of playhouses. In obedience to the law the Bath Theatre was closed, and shortly after the property was sold to the promoters of the General Hospital, who cleared it away to make room for the benevolent institution they had in view.

The amusement, however, did not become extinct with the Theatre. The proprietor of Simpson's Assembly Room had a spacious cellar under a portion of the building, and, at the risk of any legal consequences, he resolved to adapt it to Thespian uses. His enterprise was successful; he was not molested by the authorities, and it paid him well, although the performances, conducted as they were in a restricted subterranean area, were full of make-shifts, which sometimes made tragedy farcical. Neither were there any class divisions; " persons of the first quality and those of the lowest rank sat on the same bench together." Intolerable to modern ideas as was this state of things, it lasted ten years before any improvement was attempted. In the meantime, theatrical performances took place regularly, not only at Simpson's, but in "the great room at the Globe, without Westgate," the prices for admission being 2s. 6d. and 1s. 6d.; in a "large room at the George, near the Cross Bath; " also at a "New Theatre in Kingsmead Street. Pit, 1s. 6d., gallery, 1s.," the latter probably being substituted for the great room at the Globe. These undertakings testify to the great popularity of the Drama, and showed the desirability of having a properly constructed building for its display. Accordingly, a prospectus was issued in 1747 for the erection of a commodious Theatre near the Grand Parade, the money to be raised by seventy shares of £20 each. The subscribers were guaranteed not less than £3 10s. per share, and not more than £5, and they were likewise to have a silver ticket to admit to all performances, save those on benefit nights. The pioneer of progress was John Hippisley, an actor from London. He had, in fact, obtained from John Wood, the architect, the lease of a piece of land on the east side of Orchard Street for the erection of the building, for which Wood prepared the plans. The scheme met with great opposition from the vested interests threatened; but it received sufficient encouragement to warrant Hippisley, in conjunction with Mr. Roger Watts, of Bristol, to commence building. The former died soon after, and the project was in danger of collapsing, when Mr. Palmer, a brewer, came to the rescue. He and nine others provided a fund sufficient to raise a theatre, but with a less ornate elevation than Wood intended. They appear to have taken the precaution of coming, as they thought, to an understanding with Simpson, who was to close his theatre when the new one was ready for use. On its completion Simpson gave dramatic entertainments just as before, and a sharp rivalry was maintained between the two establishments. Palmer's colleagues were alarmed at the prospect, and disposed of their interest in the venture to Palmer, who, in consideration of a payment of £200 a year, induced Simpson to close his room. In 1766 Palmer, senior, retired in favour of his son, and the latter resolved to make extensive improvements in the house. While these were in progress, John Lee, an actor of merit, and father of Harriet and Sophia Lee, agreed, with two others, to purchase a piece of land near the Westgate, and to build a Theatre thereon at a cost of at least £2,500. The trio failed to fulfil the contract, and long Chancery proceedings ensued, which resulted in the defendants being amerced in damages. Free from the competition he dreaded from this quarter

Palmer, with John Lee as his manager, conducted his theatrical business with great spirit, having strengthened his position by obtaining a patent from the Crown, empowering him to establish a Theatre in Bath, with the prefix "royal"—a privilege he was the first to obtain. It has ever since been known as the "Theatre Royal, Bath."

In 1775, the house was again considerably improved; but its condition did not satisfy Palmer. He was desirous of enlarging the building, so as to make it more commodious and better suited to the rank and fashion by whom it was patronised. The proprietors were averse to his plans, and the expenditure these involved. The lessee was equal to the emergency. He caused a prospectus to be issued for erecting a Theatre, Assembly Room, and a Hotel on land in Walcot Street, the estimated outlay being thirty thousand guineas, which was to be raised on the tontine system. The "coup de main" was a decided success, as may be gathered from the following official announcement: —"The proprietors of our Theatre Royal have behaved in a most liberal manner to Mr. Palmer. On expressing his wishes that he might have it on such terms as would enable him to give the best entertainments to the city, and likewise to make an entire alteration in the building—enlarge the lobby, build an elegant tea room, scene room, etc.—they have granted him a lease for twenty years at £200 a year, he paying ground rent and taxes, and, further to encourage him, they have given up the first three years' rent towards the expenses." The grand scheme above mentioned was heard of no more. At this time Palmer annexed the Bristol Theatre, and the two establishments were worked by him with great advantage together. From the foregoing it will be seen that although the elder Palmer was at one time sole possessor of the Theatre, either he or his son disposed of it to a company or syndicate, and that Palmer, jun., was simply a lessee under it. The transference took place probably when the house was practically rebuilt in 1766, as it is styled "our new Theatre." The architect was Palmer's namesake, John Palmer, who made the subsequent alterations. The actors were known as "The Bath Company of Comedians." While the reconstruction was in hand performances were given at Simpson's Theatre, where on one occasion a remarkable contest was witnessed between the audience and the manager. Mr. Sherriffe was down in the bill to play the title part in "Richard III.," but when the curtain drew up it was found that another actor had been substituted. The piece was not allowed to proceed, and the appearance of the manager was demanded. As he refused to yield to the wish of the house, the uproar increased, and he had to beat a hasty retreat. For about an hour the suspense lasted. At length the town carried the day. Sherriffe, who was present, intimated that he would resume the rôle to oblige the audience, but not to oblige the manager. His acting that night met with tumultuous applause. As usual, the incident gave rise to sundry pasquinades.

When Mr. Palmer was devoting his attention to the improvement of the stage coaches his theatrical interests were confided to Mr. Keasbury, who in 1781 succeeded John Lee as manager. On the death of Keasbury, William Dimond, a clever actor, bought a share of Palmer's patent, and had the control of the business. The house in Orchard Street was found to be too small, and lacking the conveniences necessary for a satisfactory maintenance of the drama. A new theatre was promptly erected by Dimond (with the assistance of the same architect, John Palmer) in Beauford Square, and on its opening in 1805 the doors of its predecessor were closed.

But it was in the old house that the drama in Bath reached its meridian splendour. The reputation it had gained for encouraging young artists made it a kind of Mecca, to which aspirants for histrionic distinction eagerly directed their steps: It was well known that the approbation of the Bath playgoers was the best credential that a young actor could present to gain access to the Metropolitan boards. The weight it carried was due to the fact that the audience here was composed of the "elite" of society, gathered from all parts of the Kingdom. It was an educated body, possessing, as a rule, a wide experience of life under different conditions, and therefore qualified to judge accurately of the merits of the candidates for public favour who came before it. Very different in its composition was it to the promiscuous assemblages that filled the auditorium of other theatres. The title of "nursery of the stage" which the Bath Theatre at this time received is justified by the list of performers who hence started on their successful careers. It includes the names of Edwin, Henderson, King, Elliston, Blisset, Dimond, Abington, Crawford, Siddons, etc. The latter lady made her first appearance in Bath in October, 1778, when she played the part of Lady Townley in "The Provoked Husband." It was in the following month that her tragic power was first recognised in the rôle of Elvira in the play of "Percy." She continued to be the chief star until May, 1782, her last appearance being in a new drama called "The Distressed Mother," which was heralded by the announcement that Mrs. Siddons would produce to the audience three reasons for quitting the Bath Theatre. As her intended departure for London caused universal regret, curiosity was aroused to know the reasons actuating the favourite. These were supplied when, in the middle of an address, written for the occasion, her three children were revealed, the mother exclaiming :—

> Stand forth, ye elves, and plead your mother's cause,
> Ye little magnets—whose strong influence draws
> Me from this point, where every gentle breeze
> Wafted my bark to happiness and ease,
> Send me advent'rous on a larger main,
> In hopes that you may profit by my gain.

This little bit of maternal claptrap, as well as the pathos she threw into her leave-taking, took the house by storm. She no doubt was a fond and exemplary mother, but not so indulgent to her brother John, who was an inferior actor to herself, and who was given a dinner on his retirement from the stage. Contrasting her own lack of recognition under the like circumstances, she remarked, "Well, perhaps, in the next world, women will be more valued than they are in this."

AN OLD BATH WATCHBOX.

CHAPTER XI.

CHURCH AND DISSENT.

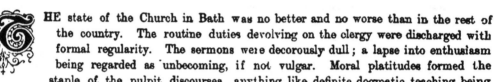HE state of the Church in Bath was no better and no worse than in the rest of the country. The routine duties devolving on the clergy were discharged with formal regularity. The sermons were decorously dull; a lapse into enthusiasm being regarded as unbecoming, if not vulgar. Moral platitudes formed the staple of the pulpit discourses, anything like definite dogmatic teaching being deemed supererogatory, as the Prayer Book contained all the enlightenment necessary on this point. Neither was the respectability of the congregations sacrificed by any undue provision being made for the poor; a bench at the back of the pews in the gallery or forms in the rear of the nave or in some other obscure position being allotted for their accommodation. Systematic pastoral visitation was unknown; when any of the humbler parishioners were " in extremis " possibly the rector or a curate would, if requested, attend to administer the last consolations of religion. None the less, the rights and privileges of the Church in temporal matters were jealously maintained and defended, as well as the formal orthodoxy bound up with it. With this narrow spirit prevailing, no wonder that the members of the Establishment were oblivious of the duty resting on them to meet the spiritual needs of the rapidly-growing population, or to save from destruction ancient fanes that were decaying or being profaned before their eyes. The most important were St. Mary de Stalls, the site of which was the south-west corner of Cheap Street, St. Michael intra muros, which adjoined St. John's Hospital, and St. Mary's, near the North-gate. These churches, built as they were at a time when ecclesiastical architecture was at or near its zenith, were stamped with the striking features of which examples survive in similar edifices dating from the same period. The solidity of the Norman and the graceful attributes of Early English were doubtless to be seen in their interiors, making them shrines worthy the high and holy purposes to which they were dedicated; but their beauty and sacredness pleaded in vain for their preservation.

With the work of rebuilding the Abbey completed, the rest of the churches within the city were regarded as superfluities or encumbrances that might with profit be dispensed with. St. Mary de Stalls, where the Mayor and Corporation were wont to worship in old time, was allowed to dilapidate; its tower when toppling was removed, and the nave soon after fell in with a crash. The other St. Mary's having been (as previously stated) secularised, the nave for the Grammar School and the tower for a prison, was demolished to give a better access to the Pulteney property. St. Michael's likewise first lost its tower, and its nave was then converted into the post office, after having been used as an ale-house, of course under the auspices of the Corporation It was probably included in the other property purchased in this locality in 1726 by the Duke of Chandos, and a roomy house soon replaced the old nave. No trace of the church is known to exist. Its memory is preserved. in St. Michael's Place, and the enclosed verdant plot facing Chandos Buildings may mark its litton or burial ground. In an antiquarian point of view at least the disappearance of these churches, small though they must have been, is to be regretted, inasmuch as, if preserved, they would have enriched the city with interesting structures of the mediæval period, the entire absence of any of which makes it in this respect contrast unfavourably with Bristol and other ancient towns. Unfortunately, whatever endowments they possessed (except in the case of St. Michael's, which was annexed to St. John's Hospital) disappeared amidst the spoliation which benefices and religious houses suffered at the Reformation; and though two others were attached to the Rectory of Bath, the resources of the latter were so limited from the same cause, that it was unable to succour the decaying members of its own household. In fact, the endowment of the Abbey was only £200 a year, and as curates had to be provided for St. Michael's extra muros, St. James' and Lyncombe and Widcombe, collections were made from strangers to eke out the receipts from pew rents to pay their stipends. *

Very different in style were the churches and chapels for which the century was responsible. Three were rebuilt and one new one added, but they were far from being "things of beauty;" on the contrary, they were and are typical of the low ebb at which Church life was running. The condition in which the Abbey was allowed to remain is evidence of the same fact. Profaned and injured as it was by the medley of houses standing against it, its degradation was regarded with indifference. No protest was heard and no effort was made to clear away these excrescences. Little was done in the interior beyond crowding it with the dead and garnishing its pillars and walls with all kinds of sepulchral monuments. The chief change, beyond the gift of the reredos by General Wade, was to have, in 1700, the six bells in the tower recast into eight bells by Abraham Rudhall, of Gloucester, "the cost to be borne by a rate raysed on the whole city of Bath"; to add two other bells in 1774, of which the Mayor, Francis Bennett, was the donor; and to provide, in 1708, a fine-toned organ by Abraham Jourdin. An extraneous but curious incident, illustrative of the intense party feeling the American War of Independence engendered, may here be mentioned. In October, 1780, a Gloucestershire clergyman obtained permission to preach at the Abbey. His text was, "Pray for the peace of Jerusalem." In unfolding his theme he denounced the war as cruel and unjust, and reviled the Government in language so displeasing to some members of the congregation that they rose from their seats and left the church. Undaunted, the minister continued his tirade until the organist was directed to play a voluntary loud and deep. Mr. Tiley, who had succeeded Chilcott in that post, obeyed with alacrity, and a roaring diapason drowned the voice of the preacher, who hastily descended from his citadel, leaving the congregation to depart without receiving the customary benediction.

---

* Not only were collections made every season among visitors towards paying the curates of the Abbey, but for the support of the Mineral Water Hospital, the Pauper Charity, and the Blue Coat School. It was likewise the custom for every family to make the pumper, the sergeants of the Baths, the attendants of the Assembly Rooms, with the servants at the house where they lodged, as well as those of the inns where their horses were stabled, a gratuity proportioned to the services rendered. Another class had also to be remembered; these were the "pious loafers," who undertook to pray devoutly for the safe journey of the departing guests, their sanctimonious professions being followed by a speedy adjournment to the gin shop or ale house to liquidate the alms received. Evidently it was an age of "perquisites," the blackmail levied being almost as oppressive as it is in the present day on board ocean-going steamers.

The first parish church whose serious dilapidations aroused attention was St. James. In 1716-17 it was renovated and enlarged, and a fresh tower added. In 1766 the nave was rebuilt, the architect being Palmer, who adopted the Italian form, although the tower was Gothic. Nor was this the only incongruity; affixed to the tower on the west side was a gabled house with a shop, the rent of which helped the income of the church. The funds for the improvement were raised by voluntary contributions and by moneys advanced on the security of the rents and Church rates, the same being liquidated by the creation of annuities. St. Michael's Church was allowed to dilapidate in the same manner; its shameful condition at length induced the parishoners to bestir themselves to repair or rebuild. The latter course was resolved upon in 1731, and Wood, having prepared plans for an edifice in the classic style, offered to erect it at his own cost, provided pews were reserved for his tenants in Queen Square. The proposal was declined, and a design by John Harvey, a stone cutter, was adopted. The cost was met by voluntary subscriptions, a rate on the whole parish, and a donation of five hundred guineas from General Wade. It took some eight years to complete the nave, and fifteen years more to build the tower. Wood, who was not an unbiassed critic, called it a "whimsical fabric," and states that the workmen, to mortify Harvey, "declared that a horse, accustomed to the sight of good buildings, was so frightened at the odd appearance of the church that he would not go by it till he was hoodwinked." Walcot, like St. Michael's, was originally a neat Gothic structure similar to those to be seen in so many country villages. It was pulled down in 1780, and replaced by an edifice like St. James, Palmer being in this case also the architect. The growth of the population compelled an enlargement of the nave, which made the tower out of proportion to it.

In the last decade of the century a praiseworthy attempt was made to provide church accommodation for the working class and poor. It was initiated by Archdeacon Daubeny, who offered anonymously £500 towards erecting a commodious building for Divine worship in which the bulk of the sittings should be free. Aided by a grant of £500 from the Christian Knowledge Society and the gift of a site at Montpelier by the Rivers family, the work was taken in hand. The foundation stone was laid on the 13th May, 1795, and the edifice was consecrated the following year. Christ Church is severely plain; but its erection fulfilled the expectations of the promoters, and removed from the Church the reproach of neglecting the poor.

The enlargements of the old churches named and the new one built did nothing to meet the spiritual needs of the dwellers of the new town which Wood had called into being. Wealthy as the occupants of this area were, much could have been done by them for the erection of one or more suitable places of worship for their own use. It is significant, however, of the apathetic condition of the Church that no attempt was made to arouse them to a sense of their responsibility to grapple with the religious exigency confronting them, more particularly as the absence of adequate endowment in the existing churches precluded help in that direction. As poverty restrained official action and ignorance of their duty made Churchmen passive, the dearth of accommodation for public worship would probably have grown into a scandal had not the scheme of erecting proprietary chapels been launched. The fact that they were commercial speculations designed to enrich individuals as well as to provide for the spiritual necessities of the population, touched not the susceptibilities of the devout any more than did the cheap utilitarian aspect that the buildings, as a rule, bore. In the case of the Octagon Chapel, Margaret's Chapel, and

Laura Chapel, houses hid them from view. Were the conditions under which they were erected not known, the inference would be that the Churchmen of the period were afflicted with the modesty that courts obscurity, or that, conscious of the meanness of these sanctuaries dedicated to the worship of the Most High, had them purposely concealed from the public gaze. The real explanation is, of course, lack of healthy, vigorous life in the Church and the dependence upon it for all sacred privileges fostered in the minds of the laity.

The originator of the Proprietary Chapel was the indefatigable John Wood. The rebuff he received over the building of St. Michael's Church stimulated him to erect a place of worship at the south-west corner of Queen Square. There were twelve shareholders, and the scheme had the approbation of the Lord Bishop of London. The chapel, dedicated to the Virgin Mary, was a stately edifice with a Doric portico, the interior being in the Ionic order. It was opened on the 25th December, 1734, and as the congregation was large and opulent, the investment proved a very satisfactory one to the syndicate. Nevertheless, it was not until 1767 that a second venture of the like character was essayed. This was the Octagon Chapel, Milsom Street, which was ready for use in that year. The next was Margaret's Chapel, Brock Street, which was opened in 1773, the first sermon being preached by the Rev. Dr. Dodd, who was, a little later, guilty of forgery, for which he underwent the extreme penalty of the law. All Saints' Chapel, an isolated building of some architectural pretensions, followed in 1790; Kensington Chapel, 1795, and Laura Chapel, 1796. The commercial principle was the basis of all; but this did not prevent them being for many years well filled with congregations composed of the "elite" of society. In addition to this select character, which made them fashionable, the pulpits were not seldom filled by able preachers, according to the standard then current for judging homiletical oratory. The emoluments they received were far better than the stipends enjoyed by curates-in-charge, who had parochial responsibilities resting on them in addition to their other ministerial duties. Of the Proprietary Chapels mentioned only two are now used as places of worship. Objectionable as they were, from the mercenary element lying at their foundation, and their common-place aspect, it may be said in their favour that they supplied a want (if not in a commendable fashion) felt throughout a large and important district. Members of the Establishment living to the west of a line drawn from St. Michael's Church to Walcot Church found these chapels serviceable and convenient until churches were erected worthy, both internally and externally, the sacred uses to which they are dedicated.

DISSENT.

It is surprising that the Church was not aroused from its lethargy by the growth of Nonconformity and the chapels it was enabled to erect by relying mainly on the voluntary gifts of its members. To this object lesson Churchmen or their pastors closed their eyes. They went on levying Church rates for the maintenance of the fabrics and other objects, which all had to pay, but no disposition was shown, as we have seen, to extend and improve the Church's ministrations by the individual liberality of those who ranged themselves under its aegis, and who formed the most numerous and wealthy section of the community. When with this state of things is contrasted the zeal, earnestness, and self-denial of the Dissenters, what wonder that sects waxed and that the influence of the Church waned? As late as 1760 there were only three meeting houses, namely, that of the Presbyterians in Frog Lane (now New Bond Street), the Baptists in Horse Street, and the Friends at the upper end of Marchant's Court, High Street. Mass was celebrated by the Roman Catholics at Bell Tree House, Bell Tree Lane, formerly the Rectory of St. James. Although the Methodists are not named, there was a small flock holding services regularly, if too obscure to be noticed. John Wesley was preaching here in 1739, when he administered a well-merited rebuke to Beau Nash, who had scoffed at religion; and often subsequently, the services being held in the open air. Though the attendances at these were large, the converts were few; in 1758 the number of adherents was only about twenty, but a couple of years later they were strong enough to rear a Bethel in Avon Street—then a respectable quarter, very different to what it is at the present time. The building becoming inadequate, the congregation removed to New King Street Chapel in 1777, the foundation-stone of which was laid by Wesley, who likewise officiated at its opening. The whole of the seats in the body of the building were free, those in the galleries only being let—a plan that was followed when Christ Church was raised. The Presbyterians, or Unitarians as they were afterwards called, having to vacate their tabernacle in Frog Lane, erected, in 1795, the existing building in Trim Street at a cost of about £2,500.

The first Baptist congregation was started in 1726 by Henry Dolling at his house near Widcombe Old Church, he having obtained a license from the Bishop's Court at Wells. It ran thus:—"The dwelling-house of Mr. H. Dolling, Widcombe, is by this court lycensed and allowed for an assembly or congregation for religious worship, as law directs." Three of the worshippers from this little community—Messrs. R. Parsons (a nephew of Dolling), Singer, and Hathaway—took a piece of land in the rear of Southgate Street (Corn Street side) in 1752, and built a small chapel, where Mr. Parsons officiated. It was opened in 1754, and the following year Whitefield preached in it several times "to a vast concourse of people." The denomination grew rapidly, and in 1768 a larger chapel was erected and opened for its accommodation in Garrard (now Somerset) Street. Mr. Parsons' pastorate extended over forty years, and during the whole of that time he received no remuneration for his services, gaining a livelihood by working at his trade of stone-carving. The Roman Catholics, abandoning Bell Tree House, constructed a chapel in St. James's Passage in 1778, which was destroyed during the Gordon riots (described in the next chapter), and then another in Corn Street in 1785, now an episcopal place of worship. The Friends transferred themselves to a chapel in St. James's Passage, the one built by the Roman Catholics, but re-edified by the Independents, who worshipped therein for a time. In 1765 the Moravians were located in their Ebenezer in Monmouth Street, which the Brethren had built.

In the same year Lady Huntingdon's chapel was opened in Harlequin Row (now the Vineyards), and for some years occupied a peculiar position. The Countess, like Wesley, wished to remain within the pale of the Church of England, and not to form a distinct sect. Strong Calvinist as she was, several beneficed clergymen became her chaplains and preached in her chapels here and elsewhere. Among them were Romaine, Venn, John Berridge, as well as Rowland Hill and Whitefield. Her aim was to leaven the upper classes of society with the evangelical truths of which she was an ardent champion. Many distinguished people attended the services at the Vineyards chapel; some criticised, others were impressed. The Earl of Buchan was one of the latter, and at his death in 1767 the corpse lay in state at the chapel for three days, funeral sermons being preached by Whitefield and others to crowded audiences. The right claimed by the Countess, as a peeress of the realm, to employ her chaplains at any time and place she thought fit, was challenged in the

Consistorial Court, London, and the judgment being against her, she had to find shelter under the Toleration Act; her ministers were obliged to take the oath of allegiance as Dissenting ministers, and her chapels to be registered as Dissenting places of worship. In consequence the parochial clergy, who were her chaplains, withdrew from the Connexion.

Notwithstanding her piety and benevolence, the Countess ruled like a female pope, particularly where the doctrine of predestination was concerned. None could be admitted or remain in the fold who did not accept it. Her views on this subject caused a rupture between her and John Wesley, whom she accused of having, at the Conference in London in 1770, denounced Calvinism and upheld justification by works. Prior to the Conference in Bristol the following year she urged her friends to attend and get the objectionable minute expunged. She also wrote to Charles Wesley, asking him to be present "to protest against such dreadful heresy." When the Conference took place mutual explanations were offered, and a reconciliation was effected. The truce did not last long; the controversy between Arminians and Calvinists broke out afresh, and was conducted with a virulence such as would have disgraced any secular discussion. Rowland Hill and some others were found not to be sound on the Calvinistic dogma, and they were forbidden to preach any more at Vineyards or in any of her ladyship's chapels. A few members of the congregation of the former, resenting this arbitrary proceeding, seceded and hired a room or outbuilding for worship in Morford Street, which was called "The Tabernacle," the first sermon being preached by the Rev. Rowland Hill in June, 1783. The little flock were not, however, allowed to remain in peace. Their separation was not forgiven, as they found when the managers of Vineyards Chapel purchased the Tabernacle over their heads and turned them adrift. In the sequel they had no cause to regret this harsh treatment. Their cause evoked the sympathy of Mr. George Welch, a banker of London and a visitor to Bath, and, aided by him, the seceders obtained the ruinated Roman Catholic Chapel, restored it, and worshipped therein under the pastorate of the Rev. Thomas Tuppen. They were called indifferently "The followers of Whitefield" and "Independents." The congregation speedily outgrew the building, and at the end of four years (1789) they removed to Argyle Chapel, which had been built specially for their use, and where the Rev. William Jay subsequently ministered for 62 years.

CHAPTER XII.

TUMULTUOUS OUTBREAKS.

HE destructive fanatical outburst in London, known as the Gordon Riots, was enacted in Bath, and though on a lesser scale, the incidents connected with it are deplorable enough to make the ebullition the saddest locally of the century. On June 9th, 1786, some lads and roughs, as evening advanced, collected together, and, headed by one John Butler, a footman to Mr. Baldwin, of the Royal Crescent, marched through the streets shouting " No Popery." The ranks of the rowdies quickly swelling, some of them rushed to the Roman Catholic Chapel (which, just completed, was ready for consecration) and commenced breaking the windows. Growing more bold, they forced open the doors, and soon reduced the interior to a complete wreck. All that was portable was conveyed to St. James's Parade and there committed to the flames. In the meantime a house adjoining the chapel, tenanted by the Rev. Dr. Brewer, the priest in charge of the Mission, was seized by the mob, who stripped it of its contents—furniture, linen, and books—and burnt them also on the Parade; they only paused in their raid to consume all the wines and other liquors found in the cellar. Dr. Brewer had a very narrow escape. He succeeded in quitting the house, but was detected and chased by the miscreants. Great as was his peril, he was refused admission to two of the principal inns, where he turned to find refuge, and was driven away by the beadles when he approached the Guildhall. He, however, contrived to slip unseen into the White Lion, and thence by a back door made his way over the river. The tumult caused the magistrates to bestir themselves, and the Riot Act was read; but the rioters were not to be deprived of their sport by the authorities. Major Molesworth, with a few Volunteers hastily collected, then arrived, and forcing an entrance to the chapel, prevented it being set on fire, which was frequently attempted. The rabble refused to disperse, and towards midnight the drum beat to arms, when about twenty more of the Volunteers, headed by Captain Duperré, marched to the scene of action with fixed bayonets, but their muskets were unloaded. Through a shower of brickbats, logs of wood, and firebrands, they struggled into the building, despite, too, the injuries some of them received. At this juncture one of the incendiaries was killed by a pistol fired, it is supposed, at one of the officers. The mob, thinking the fatal shot came from one of the Volunteers, attacked them with such fury that they had to retreat, which, it is said, they did in good order. The success of the enemy still further alarmed the magistrates, who sent expresses to Wells, Devizes, and other towns requesting the commanding officers there to come to the assistance of the civil power. The appalling spectacle the Borough Walls presented was by no means calculated to lessen the apprehension of the authorities and the peaceably-disposed inhabitants. The chapel, the priest's house, and four tenements adjoining were one mass of flames, which diffused a lurid light over the whole of the city.

The appeal for extraneous military aid was responded to with alacrity. Major Mattock, of the Queen's 2nd Regiment of Dragoons, with 40 horse, arrived from Devizes before 5 o'clock, and Captain Taylor, of the same regiment, with 60 more troopers, reached here two hours later; at 9 o'clock, Captain Barnaby, with 240 of the Hereford Militia, marched into the city from Wells. Before the arrival of these forces the main body of the rioters had dispersed, and comparative quiet prevailed. Most of the Corporation, it is recorded, "stayed up the whole of the night to watch the city and receive the officers, whom they very properly invited to an elegant dinner." It was thought at the time that emissaries from London had fomented the disturbance, and innkeepers and lodginghouse-keepers were instructed to keep an eye upon any suspicious persons, while the inhabitants generally were exhorted to detain their servants and apprentices at home. The City Volunteers, with detachments of the Regulars, horse and foot, patrolled the streets at night, and 400 special constables were sworn in; but happily there was no recrudescence of the disturbance; tranquillity reigned, and the Chamber voted one hundred guineas for distribution among the Dragoons and Militia for their promptitude in restoring the peace of the city.

Several arrests were made, including John Butler, and at the Wells Assizes, in the following August, the trials came on. Butler was sentenced to death. The other prisoners were discharged, the evidence against them being very contradictory. Butler, who was 26 years of age, was hanged on a gallows erected at the top of St. James's Parade, in the presence of a great crowd of spectators. He protested his innocence to the last, and left a written paper in which he swore before God that on the 9th of June he returned to his master's house at a quarter after nine and never went out again until eleven the next morning—an assertion which, if he could have proved, would have secured his acquittal. It is stated that neither the Scots Greys nor the City Volunteers attended the execution, but they were drawn up in Queen Square, properly accoutred, to check any disturbance had such arisen. The execution of Butler was not the closing act of this civic drama. At the Taunton Assize, in March, 1781, the Hundred of Bath was sued by Dr. Brewer for the damage sustained to his property, and he was awarded £3,754 19s. 6d. The sum was a large one, but it should be remembered that the chapel was new and was to have been opened on the 11th June; and that Dr. Brewer's house was elegantly furnished, as it was thought Lord Arundel would make it his occasional abode. Out of the amount above received a Roman Catholic Chapel in Corn Street was built.

FOOD RIOTS.

Fourteen years earlier the authorities and all lovers of quiet suffered great perturbation owing to the outrages committed in some of the surrounding towns and villages by mobs of half-starving people and idle vagabonds who found in public disorder an opportunity not to be neglected. The cause of these disturbances was the high price of provisions, consequent on a bad harvest. Ignorant of the economic laws which determined prices, the lower classes imagined that they were the victims of the greed of the farmers, corn merchants, and other retailers of food, upon whom they vented their indignation. All over the country rioting took place; corn mills were destroyed, bread and flour were seized and given away or sold at a reduced price. How general were these maraudings in this district may be gathered from the fact that the editor of a local paper announced that if he gave all the accounts he had received they would fill the whole of his columns. As a sample, the following may be quoted:—A company of women at Beckington seized the farmers' butter and sold the best at 5d. and the whey at 4d. per lb., the proceeds of the sale being handed to the owners. A mob of people at Frome compelled the hucksters to bring all the butter they had in the shops, which was disposed of at the above rates. It is not said in this case who had the cash. At the same time a load of flour on its way to the bakers was seized and divided by the looters among themselves and other poor people of the town. From one higgler they took a couple of pigs, which they cut up and distributed the joints. Similar raids at Bradford being feared, a committee of gentlemen came to an arrangement to pay the bakers of the town a bonus in consideration of their selling bread at one penny per pound. At Wincanton, to quell a riot, the leading inhabitants induced the farmers to sell wheat at a moderate price, the difference being paid by the interveners. A gang of the raiders forced themselves into the house of Farmer Bethell, at Seymour's, near Beckington. They were given food and as much beer as they liked to have in the hope of pacifying them. Instead, when fortified by the refreshments, they ransacked the house, stables, and barn. Having found four hundredweight of cheese, they appropriated it, as they did the wearing apparel and a silver watch; and further showed their senseless fury by destroying a great part of the furniture. On taking their departure the rogues carried off Mr. Bethell as a hostage, only releasing him when some miles from his house.

With these outrages occurring around the city the authorities had good cause to feel anxious for the maintenance of order within their own jurisdiction. Fortunately, the large influx of visitors gave a measure of prosperity to the toilers, and thus secured them the means of meeting the extra cost of living. A turbulent element, nevertheless, existed here, as in all other towns, which might become infected with the predatory tastes manifested elsewhere. To show that they were not indifferent to the sufferings of the poor, the Corporation, acting under a proclamation issued by the Government, announced its determination to enforce rigorously the by-laws against forestallers, regrators and engrossers of corn. The revival of this antiquated remedy was most unwise, because it tended to confirm the impression that the high prices were due, not to natural causes, but to tricks of trade. The proclamation may have helped to allay the excitement by the assurance it gave that the duty of providing means of amelioration was duly recognised by the State; but it failed, as it was bound to do, to alleviate the hard conditions prevailing. Parliament, when it met, alive to the impotency of the measures taken, hastily passed a Bill prohibiting the exportation of corn and meal and ordering all vessels laden with these commodities to be detained, also forbidding the use of wheat in distilleries. In the meantime, the Judges held special commissions to try the rioters who filled the gaols, in addition to whom several of their accomplices had been killed in conflict with the military. The leaders found guilty were condemned to death; but most of them were afterwards reprieved; some, however, were sent off to the plantations, a few were hanged, and the rest received a free pardon. During the lamentable occurrences mentioned the peace of Bath was not disturbed, although the exaggerated reports circulated caused something of a panic. The magistrates were ostentatiously vigilant in examining and testing the provisions sold; but honesty (they were proud to proclaim) ruled among the trading community. The only offender whose case is recorded was a woman who sold Yorkshire muffins, etc., about the streets. Among her stock-in-trade were found two loaves named " brown Georges" ("fancy bread" the prudent now call them) several ounces under weight. She was taken before the Justices, and, as a terror to others, was fined five pounds. As she had not wherewith to pay, she was sent to gaol, to herd, probably, with the poor debtors, and, like them, to wait until friends released her or some philanthropic individual satisfied the claims of all the creditors and secured the deliverance of those under detention. Instances of such compassion are occasionally chronicled.

CHAPTER XIII.

PHILANTHROPHIC & EDUCATIONAL.

FRIVOLOUS as was the age in many respects, the city is indebted to it for two of its most important philanthropic institutions. The first of these is the Royal Mineral Water Hospital, called formerly the General Hospital. It was provided to relieve the necessitous who came here properly accredited from any part of the Kingdom in the hope of obtaining relief from maladies through the use of the hot mineral waters. A great and increasing evil from which the city suffered was the constant influx of the maimed, the halt, the blind, and the indigent, who pestered the inhabitants and visitors for alms with the same pertinacity as the like miserable objects do in Rome and other Italian cities at the present day. This disagreeable experience was partly due to a statute of Queen Elizabeth, which gave the diseased and impotent a right to the free use of the Bath waters, justices of the peace being empowered to license them to travel to our Pool of Bethesda. The effect of the privilege was to make Bath a kind of Mecca to which indolent knaves and vagrant impostcrs made a regular pilgrimage, as well as the sick and suffering. "Beggars of Bath" became a by-word. Old Fuller, in his "Worthies," thus speaks of them : "Many were in that place ; some natives there ; others repairing thither from all parts of the land, the poor for alms, the pained for ease. Whither should fowl flock in a hard frost but to the barn door ?—here, all the two seasons, being the general confluence of the gentry. Indeed, laws are daily made to restrain beggars, and daily broken by the connivance of those who make them, it being impossible, when the hungry belly barks and howls around, to keep the tongue silent. And although oil of whip be the proper plaister for the cramp of laziness, yet some pity is due to impotent persons. In a word, seeing there is the Lazar's Bath in this city, I doubt not but many a good Lazarus, true object of charity, may be found therein." The Act of Parliament which helped thus to afflict the city expired in 1714 ; and two years later, Lady Elizabeth Hastings and Mr. Henry Hoare proposed the erection of a General Hospital for the reception of the afflicted poor who came hither for the waters. In the notice of John Wood the Elder mention was made of the many difficulties and delays which checked the realisation of the scheme. It was not till 1738 that the foundation stone was laid, and in the following year the undertaking received the sanction of an Act of Parliament. Its preamble is interesting as illustrating the serious nature of the evil and the severity of the method for its cure. "Whereas (it runs) several loose, idle and disorderly persons daily resort to the city of Bath and remain wandering and begging about the streets and the suburbs thereof, under the pretence of their being resident at Bath for the benefit of the medicinal waters, to the great disturbance of his Majesty's subjects resorting to the said city ; be it enacted, that the constables, tythingmen, and other peace officers, also the beadle or butler of the said Hospital, are hereby empowered and required to seize and apprehend all such persons and carry them before the Mayor, or some justice or justices of the peace for the said city, who shall, upon the oath of one sufficient witness, commit the said persons to the house of correction for any time not exceeding the space of twelve kalendar months, and be kept at hard labour and receive correction as loose, idle, and disorderly persons." Neither the erection of the Hospital nor the stringency of the law had any effect on the vagabond pilgrims, who swarmed here with the same persistency as the fashionable visitors whom they regarded as their legitimate prey.

The Hospital was opened on the 18th May, 1742, at which date the subscriptions received amounted to £8,643 10s. 9d., besides the gifts of the stone, timber, etc., used in the building. It is interesting to note that from the outset provision was made for the admission of soldiers, who were required to bring a certificate from their commanding officers stating to what corps they belonged, the same to be preceded by a description of their cases. The many cures effected soon caused the institution to be widely known, and the applicants for benefits became more numerous than the wards could accommodate. Before the end of the century (in 1792) the desirability of enlarging the Hospital was seriously broached by some of its friends, as well as its removal to another site. As these proposals were deemed unwise, it was suggested that a second Hospital should be built, or a house hired for the reception of thirty more patients, bringing the number up to 130. This course was more favourably regarded by the Governors, and on the strength of it (probably also to force their hands), the Progressives (as they may be styled) purchased the Alfred Hotel and premises for £1,300, at the same time intimating that they would retain possession of the property for two years to enable the Governors "to reflect coolly on the matter." Reflection did not convince them, and the Hotel was transferred to the Bath City Infirmary and Dispensary. The hesitation of the Governors to undertake the additional responsibility projected was justified by the difficulty of securing adequate funds to maintain the Hospital on its existing basis. In 1783, and again in 1797, they had to announce that a reduction of the beds must be made unless the public support were increased. On both these occasions there was a response sufficiently liberal to avert the threatened diminution of the benefits of the charity. How well it has flourished since is shown by the splendid additions made to the building and the improvement effected in its surroundings, in the last half of the nineteenth century.

The poor of Bath were excluded from the Hospital on the ground that their homes were contiguous to the Baths, and that other facilities for using these were within their reach. This exceptional treatment drew attention to the suffering caused to the sick poor who were unable to get prompt and efficient medical aid. The absence of any such provision induced a few benevolent gentlemen to take, in 1747, a house in Kingsmead Street and provide the professional skill and medicine required for dealing with these cases. It was known as the Pauper Charity, and evoked a considerable amount of sympathy. As stated in an earlier chapter, the proceeds of the sale of the poems contributed to the Batheaston Vase were added to its funds; concert breakfasts with cotillons were likewise given in its behalf at Simpson's Rooms. In 1788 a Casualty Hospital was opened in another house, where persons injured by accident could at once obtain the best surgical assistance and every comfort. In 1792 the two charities named were amalgamated under the title of the Bath City Dispensary and Asylum. A large house on the Lower Borough Walls, formerly a tavern (Alfred Hotel), was secured, and furnished for the reception of patients suffering from acute or contagious disases, in addition to which the sick, in some instances, were visited at their homes. The number of out-patients in 1799 was 1,536, and of in-patients 113. The title of the institution was changed in 1826 to the Royal United Hospital. The fine block of buildings now bearing the name is the outcome of the small Pauper Charity of 1747.

SCHOOLS.

While the bodily sufferings of the indigent were receiving attention, some praiseworthy efforts were and had been made to place the means of education within the reach of a portion, at least, of the needy among the rising generation. Not that mental culture was by any means common among the well-to-do; in the case of females it was generally regarded as superfluous. Swift, in one of his letters to Mrs. Delaney, remarks: "A woman of quality, who had excellent good sense, was formerly my correspondent; but she scrawled and spelt like a Wapping wench, having been brought up in a Court at a time before reading was thought of any use to a female; and I know several others of very high quality with the same defect." As early as 1711, Dean Willis (afterwards Bishop of Winchester), Robert Nelson, the author of "Fasts and Festivals," and others raised a subscription to open schools for the education and clothing of fifty boys and fifty girls, and the apprenticing of them to useful trades. During eleven years the schools were conducted in private houses, but in 1722 a spacious edifice for the accommodation of the pupils was erected on the Sawclose, at a cost of about £1,000, towards which Mr. Skrine, of Warleigh, gave £100 and ten tons of timber. A descendant of the latter—the late Mr. H. D. Skrine, of Claverton Manor—was, until his death in 1901, a generous supporter of the school. The foundation-stone bore the motto, "God's Providence is our Inheritance," a declaration which the subsequent history of the institution has not belied. Though two centuries have nearly elapsed since its foundation, it still exists and its ancient lines are maintained. After Queen Square Chapel was opened, the sacramental alms sufficed not only to relieve some of the poor, but to provide an education for twenty children. At a later period of the century the Octagon Chapel congregation supported a school for girls, who were likewise clothed at its expense. The sympathy thus shown in the work of training the young was creditable to both these places of worship, seeing how little was done in that direction.

The chief educational movement of the time was, however, the introduction of Sunday Schools, pregnant as they were with far-reaching results on the secular, as well as the religious, training of the young. The honour of initiating in Bath these valuable institutions is due to Mr. Henry Southby. Seeing that the poor had no means of learning their religious and social duties, he, in January, 1785, started a subscription for the establishment of Sunday Schools. In a few weeks a sum sufficient for commencing operations was received. At a meeting of the subscribers, the clergy of the city and several laymen were appointed a committee, and among the resolutions passed were the following: "That all children recommended from the parishes of Bath, Walcot, Bathwick, and Widcombe should be admitted into the school; that the children should attend Divine service every Sunday at the Abbey; and that the books of instruction should be such only as were in the list of those recommended by the Society for Promoting Christian Knowledge." A large house in St. James's Street was taken, and six other schools were in due course established.

In the first year a very interesting development occurred, the particulars of which may be read with pleasure by those who are trying to combine industrial and technical training with the ordinary school curriculum. At the suggestion and cost of Lady Maria and Lady Isabella Stanley, a School of Industry was opened, the premises in St. James's Street being devoted to its use, and the ladies above named acting as the superintendents. In it fifty girls, selected from the Sunday Schools, were instructed in knitting and plain work, likewise in reading. Two other schools were held in the same building for smaller children. Here twenty boys and twenty girls were employed in spinning, and "as an encouragement for their industry they have the liberty of attending a master and mistress for instruction two hours every evening." Thus "Evening Continuation Classes" were engrafted on the day schools in the early days of the educational movement. Later the children were employed in the following manner :—

Thirty little boys to knit stockings and garters and make garden nets.

Thirty little girls to knit stockings and garters.

Thirty boys and thirty girls to spin wool for cloth and worsted.

Thirty girls to spin flax for linen.

Thirty girls to sew, make the linen, wearing apparel, etc., for the scholars.

The larger girls cleaned the house and washed the clothes to enable them "to earn a decent livelihood, become good servants, and, instead of being a burden, be a benefit to society. As the girls go progressively through all the schools, they learn to make their own stockings, spin wool and linen for their clothes, as well as to make and mend them." In May, 1787, it is recorded that "last Sunday one hundred and twenty children belonging to the School of Industry were at the Abbey Church dressed in their new clothes. Their appearance gave general satisfaction, and particularly so as the clothes are the produce of their own labour. The remaining sixty children will be clothed in the same manner very soon. The apparel for the large boys consists of a coat and waistcoat of olive-coloured serge, leather breeches, shirt, shoes, and stockings, and a cap of serge, with the words 'Reward of Industry' round it. The little boys have jackets and petticoats of the same serge, shirt, shoes, stockings, linen tippet, and cap, with a piece of worsted in imitation of ribbon, inscribed with the same words as those on the other caps." At this date there were as many as seven hundred children in the Sunday Schools, and before the century closed they had reached about one thousand. From the above description it will be seen that the schools were not only of great benefit religiously and morally, but that they were conducted on a system as practical as philanthropic. Lady Stanley of Alderley (Maria Josepha Holroyd), in one of her letters dated 1787, gives an interestng picture of the Sunday School scholars when assembled for worship :—"It was last Sunday (she writes) I particularly wished to have had you witness to a scene that struck me beyond description. It was at our Cathedral, which we call the Abbey. I dare say you have heard of Sunday Schools. It is but lately we have had that institution here, and at first it went on slowly, but by joining to it a school of industry, they now all crowd to the other, which is a necessary step to that of industry. There is a clergyman employed for this Sunday evening service for the children alone, after the other common service is over, and it is in the great Isle (sic) where you must suppose nine hundred children in perfect order, placed on benches in long rows, so quiet that you could have heard a pin drop while the clergyman was reading. Reflect how very extraordinary the circumstances alone! when you recollect that most of them were taken out of the streets, untaught and actually almost savage, cursing, swearing, and fighting in the streets all day, and many without a home at night. Two girls, I myself know, slept in the street. Most of them not only ragged and starving, but without a chance of being put in the way to earn their bread. Yet there I saw them, not only in such order, but so well instructed as to have most of the service by heart; for though they had books, I observed they scarce looked at them, and yet repeated the responses perfectly aloud. At one instant also, without direction to do so, the nine hundred dropped on their knees and rose again, which showed they knew what they were about; their little hands lifted up and joined together, looking with such innocent devotion. They sang the Psalms, all in time with the organ, by heart, and notwithstanding the number, the sound was neither too loud nor too harsh, but on the contrary soft and affecting beyond measure. . . . How much this order and decency must civilise these children, and what a great step this is towards reformation of morals!" Fanny Burney when in Bath in 1791 visited the schools, and described her impressions in the following lines:—" Such a number of poor innocent children all put into a way of right, lifting up their hands and joining in prayers and supplications for mercy and grace, which, even if they understood not, must at least impress with a general idea of religion." Lady Spencer was then the patron in succession to the Ladies Stanley, and Miss Burney records that she had a lottery, without blanks, of toys and playthings for the children.

The School of Industry continued to flourish a few years longer. As in the case of so many other institutions, past and present, it was crippled in its finances, the falling off in the support it received being attributed partly to the fiscal effects on the trade and the pockets of the lieges produced by the war with France. Some of its friends urged the committee to omit the Church Catechism and to alter the rule requiring attendance at church, as Nonconformists were thereby debarred from the benefit of the charity. The relaxation suggested was strongly supported by Mr. Wilberforce, who thought it would increase the funds. The committee declined to accede to the request, and the operations of the School became more and more restricted. In the meantime, a District National School, on the plan of Dr. Bell, was formed, the boys assembling in a room in Kingsmead Square, and the girls in rooms in various parts of the city. The great success attending the scheme made concentration desirable, and the committee, after a long quest, were fortunate to obtain the Weymouth House premises from the then Marquis of Bath, on peculiarly liberal terms, and there they built a school for a thousand children. It comprised a day school for boys, a Sunday School and a library for the improvement and amusement of the children and their parents at their homes, in addition to an industrial school for fifty girls.

CHAPTER XIV.

TRADE AND COMMERCE.

HE decline of the cloth trade in Bath was accelerated, and the growth of other manufacturing industries was checked, by the fashionable character the city assumed. The steady increase in the number of visitors gave full employment to the trading and working portions of the population. Businesses grew and multiplied for satisfying the numerous wants of invalids and pleasure seekers, for whose accommodation more and more lodging-houses were needed. Thus plenty of occupation, and much that was easy, was provided for assistants, servants, and others lower in the social scale. There was, therefore, no special inducement for starting great commercial enterprises or for saving from extinction such an ancient avocation as that of cloth-making, for which Bath was once so renowned. Even when it ceased to be made here the repute of the cloth was such that at Belcomb, near Bradford, was manufactured a thin superfine cloth for the robes of state worn by the ladies of the Sultan's harem at Constantinople. At Twerton three large factories were kept going, the weavers and other employees being recruited from the workhouses in and around Bath. In one of these establishments alone there were from eighty to a hundred pauper girls and boys employed. Apprenticed when seven years of age and upwards, their servitude continued until they were twenty-one, a trifling fee being paid with each when the indenture was signed. The system was to board them out among the weavers, and what with very long hours in the factory, poor food, and often brutal treatment by their foster parents, the death-rate among these waifs, many of them orphans, was very high. It is significant of the ill-repute of the weavers that in later years, when the overseers of some of the city parishes announced that they had boys or girls to be apprenticed, the notice concluded with the intimation, " No clothworker need apply."

If, with the exception of the stone trade, so extensively developed by Ralph Allen, Bath was without great manufactories, it had, as we have mentioned, a number of minor industries, some of them affording scope for a higher class of artificers. Painters, carvers, engravers, jewellers, and gilders were numerous. Seal engraving was introduced and successfully carried on by John Wicksteed. The stone carvers became noted through their productions both far and near, and the same was the case with Bath rings and Bath metal work. A brisk trade was also done in lace; but this, it seems, was of Honiton make, the designs only being sent from Bath. According to a very ancient custom the various trades were wont to walk in procession annually. The tailors (who wore gowns), cordwainers, plasterers, masons, joiners, mercers, upholsterers, butchers, bakers, etc., paraded the streets in their regalia, headed by a band of music. The practice had, it is said, existed from the fourteenth century, but its antiquity did not deter the authorities from decreeing its abolition on the ground that there was no foundation beyond custom for its maintenance, and probably also that it was not conducive to sobriety and good order.

The most noteworthy commercial enterprises were making the river navigable between Bath and Bristol, and the construction of the Kennet and Avon Canal, the former identified with the opening of the century and the latter with its close. It was the Duke of Beaufort who first suggested the making of the Avon navigable as a means of improving the trade of Bath. His Grace testified his faith in the undertaking by obtaining in 1711 an Act of Parliament, at his own expense, authorising the work, and promising to co-operate with the Bathonians in its execution. Delays occurred, and the scheme remained in abeyance until 1722, when it was revived by Mr. John Hobbs, merchant, of Bristol, who got together a company of thirty-two shareholders to raise the capital required. Locks were formed at Saltford and Twerton to avoid the weirs at those points; the bed of the river at Newton, where it was fordable, was deepened, and a bridge carried over it to replace the ford, and other improvements effected. The first barge, laden with merchandise, reached the city from Bristol on the 15th December, 1727, the achievement being regarded with the same admiration as the making of a railway into a new district would create now. When coal was brought from Shropshire by the new route the colliers of the neighbourhood threatened to destroy the locks. To protect the property an Act of Parliament was obtained in 1735, which made it an offence punishable with death, without "benefit of clergy," to destroy any lock, or floodgate on any navigable river. If the culprits were not traced the Hundred was to pay the cost of the damage. Nevertheless, an early example of a trade outrage was furnished. In the following year the Saltford lock was destroyed, some of the stonework being blown up with gunpowder. The raiders left notices to the effect that three hundred men had done the mischief and that a thousand would carry it on if an immediate stop were not put to bringing the coals by water, avowing that they had better be hanged rather than that they and their families should be compelled to starve. The proprietors at once offered a reward of twenty pounds for the discovery of the perpetrator of any such outrage, and nothing of the kind was afterwards attempted.

The making of the Avon thus serviceable to commerce was, however, only the first step in a scheme for opening up water communication with London. As early as the reign of Elizabeth a project of this magnitude was started by a Rowland Vaughan and one Mr. Hill. The idea was welcomed in commercial circles; but the time was not ripe for its execution, and the political troubles which subsequently arose caused it to be almost forgotten. After the Civil War some London merchants took the undertaking in hand, and Cromwell offered to join them and subscribe twenty thousand pounds towards it. Surveys were made, as a result of which it was proposed, by means of a canal, to connect the Avon at its source with the Thames near Cricklade. Charles II., on the Restoration, encouraged the execution of the design, and a Bill passed the Commons authorising the work; but after being read twice in the House of Lords it was dropped. Beyond making the Avon navigable, under the auspices of the Duke of Beaufort, nothing was done until the last decade of the century, when a company was formed for the purpose of constructing the Kennet and Avon Canal. The plan, in this instance, was to make a waterway from the Avon at Bath to join the Kennet at Newbury, from which place the navigation of the Thames was open. The Act for its execution was obtained in 1795. A vigorous commencement was made, but it was soon discovered that the cost of the Canal had been under-estimated, and this paralysed operations for a time. Fresh legislative powers were obtained, and the Canal was completed in the first year of the last century.

CHAPTER XV.

ROMAN ANTIQUITIES.

N excavating for the foundation of so many new streets and building on the site of the old Roman city, it was only natural that sundry relics of antiquity should be brought to light. There were, indeed, constant finds, not only of altars and monuments, but remains of the stately edifices that once adorned the Aquæ Solis of the Latin conquerors, including their extensive Baths. The varied contents of the museum beneath the Roman Promenade and at the Literary and Scientific Institution were nearly all found in the course of the 18th century. Such, however, was the absence of anything like a reverent archæological spirit that much of surpassing interest, after being regarded as a mere curiosity, was swept away, as if of no value. The indifference thus manifested is the more remarkable seeing that all the moveable antiquities—pieces of sculpture, altars, inscriptions, fictile vessels, etc.—were preserved, and form the present unique collection of Roman antiquities. Even these were for long dispersed in various nooks and corners, obscured with dirt and exposed to constant mutilation. It was not until the Rev. Richard Warner, in 1795, obtained permission of the Corporation to purify and arrange them in a separate building that they received any attention. Warner increased the debt due to him by writing and publishing a descriptive catalogue of these antiquities.

The most important discovery was that made, as already stated, when the Abbey House, which formed a portion of the palace of John de Villula, Bishop of Bath, 1088 to 1122, was pulled down in 1755. The palace and monastery covered the ancient ruins, which probably supplied a good deal of the stone for their erection, just as in the sixteenth century the Bishop's palace, then in decay, furnished the stone for building the first Guildhall. It was fortunate that the level of the Roman city was much below its successors, otherwise there would have been very little left to tell of its magnificent buildings, or of the extensive use made of its mineral springs by these early settlers from afar. The ruins referred to were twelve feet below the surface of the ground, and consisted of a bath, running north and south, forty-three feet in length and thirty-four in breadth, included within walls eight feet in height, built with wrought stone, lined with terras, and ornamented with twelve pilasters; and, secondly, of a semi-circular bath, to the northward of the former, measuring from east to west fourteen feet four inches, and from north to south eighteen feet ten inches, with four pilasters, and containing a stone chair eighteen inches high and sixteen inches broad; thirdly, two large rooms to the eastward, with tesselated pavements, each thirty-nine feet by twenty-two, used as sudatories, having double floors, on the lower of which stood rows of pillars composed of square bricks, which sustained a second floor, formed of tiles and covered with two layers of firm cement, two inches thick; the stones and bricks bore marks of fire, and the flues were thickly charged with soot. One of the furnaces which heated these hypocausts was still visible, and at its mouth were scattered pieces of charcoal and burnt wood, testifying the use to which it had been applied. Flights of steps led to the baths; they were six inches thick, but from long use they were three inches out of the square. These rare buildings, which the earth had preserved for some fifteen hundred years, were destroyed to make way for modern houses. Some idea of the existence of a large oblong bath and other smaller ones to the west was then current; but the possibilities which this surmise opened up failed to save the eastern wing of this great balneal system. The western portion, having been uncovered of late years, is preserved, and likewise the oblong bath which connected the two wings.

Vestiges of an extensive building that once existed on the site of the present Royal Mineral Water Hospital and Blue Coat School were brought to light. In digging the foundation of the former institution Wood imagined that he had come on a part of the Prætorium. He was led to this conclusion from the finding of a hypocaustum, which he stated was also "the foundation of an altar placed near the General's tent," a number of pilæ, some nine and some six inches square, for carrying hot air to the several rooms; two mosaic pavements, one six feet broad, the other eighteen feet, the pattern of both being graceful convoluted circles; two steps of six inches rise; a floor paved with common stone; and a wall three feet two inches thick. Wheat was also found. These relics, which were effectually got rid of, were six feet below the surface of the ground, and three above the gravel or subsoil. At a later period two other tesselated pavements were brought to light to the west of Wood's "Prætorium," and probably were originally under the same roof.

In excavating on the Terrace Walks for the second Assembly Rooms, built there by Mr. Thayer in 1729, the workmen came upon an ancient place of sepulture. Some of the bodies had been buried in stone coffins hewn out of one solid block, and others in those made of several stones. Those without coffins were wrapped in hides, and had shoes on their feet; others had leather shrouds, but no shoes. The latter were doubtless monks who died in the adjoining monastery; the former date back to Saxon and Roman times. Teeth and bones of horses were found, which were sold by one of the labourers to a confiding public as belonging to men of gigantic stature.

The extent and endurance of the Roman occupation are further attested by the large number of interments in and around Bath due to this period. Twelve stone coffins were disinterred in one spot at Weston, besides several isolated ones. Six were removed from Rivers Street; in one was a coin of Constantine. At Combe Down five, with skeletons, were found in the course of making a garden for a new house, besides two stone chests, one of which contained bones and the other a horse's head; two drinking cups, embossed with hares, boars, and deer, mingled with tendrils of ivy; 230 coins, chiefly of the later Empire; many iron implements; two keys; pottery of various kinds; a sheep's bell, and a bell for the breast belt of a horse. On one of the coffins was an inscribed slab, face downwards, dedicated to Marcus Aurelius Antoninus, or Caracalla, by his freed man, Nævius. Like remains have been found at Widcombe, Wellow, Inglescombe, Farleigh, Newton, Box, Colerne, and other places, as well as tessellated pavements, or fragments of such, the principal of these being at Wellow and Newton.

Of Roman finds other than structural the chief were the beautiful bronze head of Minerva, unearthed in Stall Street in 1727, which, happily, has been preserved uninjured; and a bronze statuette, of fine design, supposed to represent the Emperor Trojan, found in

Monmouth Street. Near the large mosaic pavements still buried at Wellow the fragment of a bas relief was obtained depicting three figures, a male and two females. The latter are draped; one of them holds in her left hand a staff or forked instrument; the male holds an apple or pomegranate in his right hand.

With these gleanings in the archæological field the task of the author is done. To him the preparation of the work has been an attractive one and he trusts that it will not be uninteresting to the general reader, nor without value to the student of history. The Nineteenth Century, which has now closed, awaits similar treatment. Happily, one third of the work has been accomplished : the "Annals of Bath," published by Capt. Mainwaring, covers the period from 1800 to 1834 inclusive, and the political history of the city, down nearly to the close of 1900, has been written by a later hand. These beginnings considerably lighten the task the completion of the century's review involves. Mainwaring derived his facts mainly from local newspaper files, and the same mine of wealth is open to anyone who may like to follow the gallant captain's example. True, local history shares the fate of standard literature in being too much neglected, owing to the absorbing attention the overwhelming output from the press of periodical and light literature commands.

When the minds of the masses are really well trained, instead of the memories being gorged as now with indigestible pabulum, under a showy system of standards, we may expect a gradual increasing per-centage of readers of works calculated to enlarge the understanding, and thus give the nation a higher mental status than it at present enjoys. In view of a good time coming, local records should be kept in print well up to date, and the rising generation be encouraged to take an interest therein. With the new departure we are making in education, it should not be difficult to add to the curriculum an occasional lesson in the history of the town in which the boys and girls live. The knowledge thus gained would help to foster attachment to the place, give them a relish for historical, biographical and other mind-informing works, and in this way supply an antidote to the craving for the trashy novel and scrappy paper the universal prevalence of which is deplored by all who value the intellectual and moral well-being of the country.

THE END.

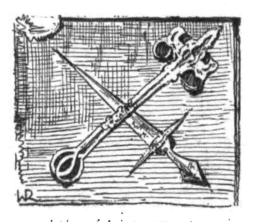

NOTES.

1. NASH'S RULES (p. 29).

By General Consent, Determined. I.—"That a visit of ceremony at coming to Bath, and another at going away, is all that is expected or desired by Ladies of Quality and Fashion—except Impertinents." II.—"That Ladies coming to the Ball appoint a Time for their Footmens coming to wait on them Home, to prevent Disturbances and Inconveniences to Themselves and Others." III.—"That Gentlemen of Fashion never appearing in a Morning before the Ladies in Gowns and Caps shew Breeding and Respect." IV.—"That no Person take it ill that any one goes to another's Play or Breakfast, and not to theirs;—except Captious by Nature." V.—"That no Gentleman give his Tickets for the Balls to any but Gentlewomen;—N.B. Unless he has none of his Acquaintance." VI.—"That Gentlemen crowding before the Ladies at the Ball, shew ill Manners; and that none do so for the Future;—except such as respect nobody but Themselves." VII.—"That no Gentleman or Lady take it ill that another Dances before them; except such as have no pretence to dance at all." VIII.—"That the Elder Ladies and Children be contented with a Second Bench at the Ball, as being past, or not come to Perfection." IX.—"That the younger Ladies take Notice how many Eyes observe them;—This don't extend to the Have-at-all's." X. "That all whisperers of Lies and Scandal be taken for their Authors." XI.—"That all Repeaters of such Lies and Scandal be shun'd by all Company;—except such as have been guilty of the same Crime." N.B. Several Men of no Character, Old Women and Young Ones, of Questioned Reputation, are great Authors of Lies in this Place, being of the "sect of LEVELLERS."

2. BATH ATTRACTIONS (p. 37).

Lady Luxborough, writing to the poet Shenstone in 1752 gives the following account of Bath:—"We can offer you friendly conversation, friendly springs, friendly rides and walks, friendly pastimes to dissipate gloomy thoughts, friendly booksellers, who for five shillings for the season will furnish you with all the new books; friendly chairmen who will carry you through storms and tempest for sixpence and seldom less, for Duchesses trudge the streets here unattended. We have also friendly Othellos, Falstaffs, Richard III.'s, and Harlequins, who entertain one daily for half the price of your Garricks, Barrys, and Richs; and (what you will scarcely believe), we can also offer you friendly solitude, for one may be an Anchorite here without being disturbed by the question "Why?" Would you see the fortunate and benevolent Mr. Allen, his fine house and his stone quarries? Would you see our lawgiver, Mr. Nash, whose white hat commands more respect and non-resistance than the crowns of some Kings? To promote society and good manners, and a coalition of parties and ranks, to suppress scandal and late hours are his views, and he succeeds rather better than his brother monarchs generally do."

3. ST. JOHN'S HOSPITAL (p. 65).

According to the preamble of Sir John Trevor's award in 1716, the Hospital had been advanced to great possessions, worth about £1,200 per annum, which the Mayors of Bath, who were for the time being Masters, disposed of as they pleased, the nominal masters being allowed a dole of 40s. per annum only.

4. ALLEN'S MEDAL (p. 80).

The Duke of Cumberland in 1752 presented Ralph Allen with a silver medal, bearing the following inscription—"The gift of his Royal Highness, W. D. of Cumberland to the famous Mr. Allen, 4 Dec., 1752." No one seems to have known anything of this act of generosity, and what is still more strange, the medal left Allen's possession a year or two after the above date, and found its way to Ireland where a Mr. John Campbell converted it into a Masonic emblem. It was picked up in a shop in Belfast by Mr. D. Buick, J.P., some twelve years ago. An engraving of it was published in the "Bath and County Graphic" for September, 1901. Its singular history raises a doubt whether it was ever presented, or, if presented, whether appreciated by the recipient.

5. THE PALMER TABLET (p. 82).

The following is the inscription on the tablet in the Abbey Church:— "Sacred to the memory of John Palmer, esq., originator of the Mail Coach System, Comptroller-General of the Post Office, and many years M.P. for this city, Obit. Aug. 16th, 1818, aged 77. And to his sons, the Rev. John Palmer, M.A., Rector of Peldon, Essex, Obit. May 17th, 1851, aged 77. Major-General Charles Palmer, formerly of the 10th Hussars, M.P., for this city, Obit. April 17th, 1851, aged 74; and Captain Edmund Palmer, R.N., C.B., Obit. September 19th, 1834, aged 52."

6. THE DRAMA (p. 105).

The statute of 1736, the provisions of which governed the position of actors for more than a century afterwards, ordained that from and after the 24th June, 1737, every person (not being legally settled in the place, or if without Letters Patent from the Crown or License from the Lord Chamberlain), who shall for hire, gain, or reward, act, represent, or perform, or cause to be acted any interlude, tragedy, comedy, opera, play, farce, or any other entertainment shall be deemed to be a rogue and vagabond and subject to all the penalties and punishment the law prescribes for this class of misdemeanants.

CPSIA information can be obtained
at www.ICGtesting.com
Printed in the USA
BVHW070259171122
652078BV00004B/18